FROM THE LIBRARY OF
R. W. SHEREMETA
ASSOCIATES

Simplified Design

REINFORCED CONCRETE BUILDINGS OF MODERATE SIZE AND HEIGHT

Edited by Gerald B. Neville

An organization of cement manufacturers to improve and extend the uses of portland cement and concrete through scientific research, engineering field work, and market development.

5420 Old Orchard Road, Skokie, Illinois 60077-4321

© 1984 Portland Cement Association

First edition

Printed in U.S.A.

Library of Congress catalog card number 84-61182

ISBN 0-89312-043-X

This publication was prepared by the Portland Cement Association for the purpose of suggesting possible ways of reducing design time in applying the provisions contained in the ACI 318-83 "Building Code Requirements for Reinforced Concrete."

Simplified design procedures stated and illustrated throughout this publication are subject to limitations of applicability. While such limitations of applicability are, to a significant extent, set forth in the text of the publication, no attempt has been made to state each and every possible limitation of applicability. Therefore, this publication is intended for use by professional personnel who are competent to evaluate the information presented herein and who are willing to accept responsibility for its proper application.

The Portland Cement Association shall not be liable for any injury, loss, claim, or damage that arises out of or is in any way connected with the use of this publication. If, notwithstanding the foregoing, PCA should be found liable for loss or damage arising out of or in any way connected with the use of this publication, PCA's liability shall be limited to fifty dollars ($50.00). The provisions of the preceding two sentences shall apply if loss or damage, irrespective of cause or origin, results directly or indirectly to person or property from negligence, active or passive, of PCA, its agents, or its employees.

Foreword

The Building Code Requirements for Reinforced Concrete (ACI 318) is an authoritative document often adopted and referenced as a design and construction standard in state and municipal building codes around the country as well as in the specifications of several federal agencies, its provisions thus becoming law. Whether ACI 318 is enforced as part of building regulations or is otherwise utilized as a voluntary consensus standard, the fact remains that design professionals use this code almost exclusively as the basis for the proper design and construction of buildings of reinforced concrete.

The ACI 318 code applies to all types of building uses; structures of all heights ranging from the very tall high-rises down to single-story buildings; facilities with large areas as well as those of nominal size; buildings having complex shapes and those primarily designed as uncomplicated boxes; and buildings requiring structurally intricate or innovative framing systems in contrast to those of more conventional or traditional systems of construction. All these technical considerations contained in general formulae that are developed to encompass the extremes of building design and construction would seemingly make the application of ACI 318 code provisions both complex and time consuming. This need not be the case, particularly in the design of reinforced concrete buildings of moderate size and height.

It is to this end, then, that this book has been written as a timesaving aid for use by experienced professionals who consistently seek ways to simplify design procedures.

This manual was prepared under the direction of Gerald B. Neville, Manager of Structural Codes, PCA Codes and Standards Department, who provided technical assistance to the individual authors when needed, and who also served as editor and coordinator for the final manuscript.

The creative work of each of the authors whose names appear under the several chapter headings is gratefully acknowledged. Our thanks also go to Herman L. Szeker, S.E., who provided an independent review of the entire contents and a computerized check of each of the building design examples using the simplified methods presented in the book in comparison to solutions obtained by using the full ACI 318 procedures.

In reading and working with the material presented in this book, design professionals are encouraged to send their comments to PCA together with any suggestions for further design simplifications. Commentaries will be appreciated and new ideas may be incorporated in the texts of future editions of this publication.

JAMES P. BARRIS
Director, Codes and Standards Department

Contents

1

A Simplified Design Approach

Gerald B. Neville*

First. . . .the Stage for Simplified Design

1.1 THE BUILDING UNIVERSE

There is little doubt that the construction of a very tall high-rise building, a large domed arena, or any other prominent megastructure attracts the interest of a great number of structural engineers around the country. The construction of such facilities usually represents the highest level of sophistication in structural design and often introduces daring new concepts and structural innovations as well as improvements in construction techniques.

Many structural engineers have the desire to become professionally involved in the design of such distinctive buildings during their careers. However, each year there are very few projects of this prestigious caliber built. Truly, the building universe consists of low-rise and small area buildings. Fig. 1-1 shows the percentage of building floor area constructed in 1983 in terms of different building height categories. From this it can be readily seen that the vast majority of the physical volume of construction is in the 1- to 3-story range.

*Manager, Structural Codes, Codes and Standards Dept., PCA

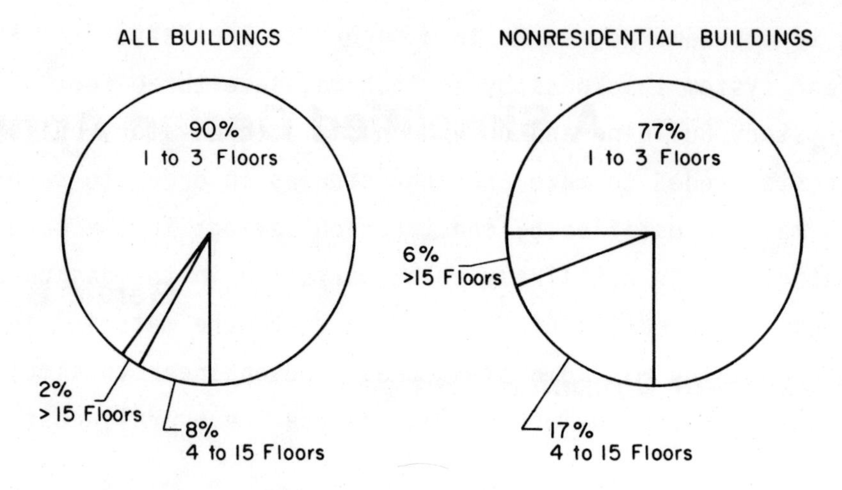

Fig. 1-1 Floor Area of Construction, 1983*

In the same way, Fig. 1-2 shows the percentage of nonresidential building projects constructed in various size categories. Building projects less than 15,000 sq ft dominate the building market.

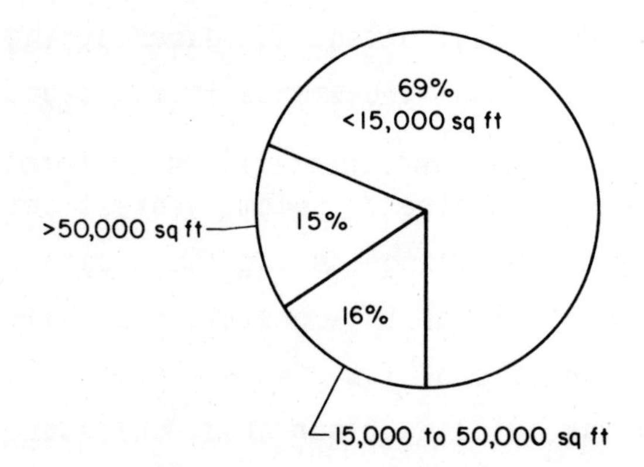

Fig. 1-2 Nonresidential Building Project Size, 1983*

When all these statistics are considered, it becomes apparent that while most engineers would like to work on prestigious and challenging large high-rise buildings or other distinctive structures, it is more likely that they will be called upon to design smaller and lower buildings.

---

*Derived from F. W. Dodge Division, McGraw-Hill Information System Company, Dodge Construction Potentials, with permission (May 1984).

1.2 COST EFFICIENCIES

The benefit of efficient material use is not sought nor realized in a low-rise building to the same degree as in a high-rise facility. For instance, reducing a floor system thickness by an inch may save three feet of building height in a 36-story building and only 3 in. in a three-story building. The added design costs needed to make thorough studies in order to save the inch of floor depth may be justified by construction savings in the case of the 36-story building, but is not likely to be warranted in the design of the smaller building. As a matter of fact, the use of more material in the case of the low-rise building may sometimes enable the engineer to simplify construction features and thereby effectively reduce the overall cost of the building.

In reviewing cost studies of several nonresidential buildings, it was also noted that the cost of a building's frame and envelope represent a smaller percentage of the total building cost in low-rise buildings as compared to high-rise structures.

In low-rise construction, designs that seek to simplify concrete formwork will probably result in more economical construction than those that seek to optimize the use of reinforcing steel and concrete, since forming represents a significant part of the total frame costs. There is less opportunity to benefit from form repetition in a low-rise building than in a high-rise building.

In considering the responsibility of the engineer for providing a safe and cost-effective solution to the needs of the building occupant and owner, it becomes clear that for the vast majority of buildings designed each year there should be an extra effort made to provide for expediency of construction rather than efficiency of structural design. Often, the extra time needed to prepare the most efficient designs with respect to structural materials is not justified by building cost or performance improvements for low-rise buildings.

1.3 THE COMPLEX CODE

In 1956 the ACI 318 Code was printed on 73 small-size pages; by 1983, the Code contained more than 100 large-size pages and over 150 pages of Commentary -- a very substantial increase in the amount of printed material with which an engineer has to become familiar in order to design a concrete building.

To find the reasons for the proliferation in code design requirements of the last twenty-five years, it is useful to examine the extensive changes in the makeup of some of the buildings that required and prompted more complex code provisions.

1.3.1 Complex Structures Require Complex Designs

Advances in the technology of structural materials and new engineering procedures have resulted in the use of concrete in a new generation of more flexible structures, dramatically different from those for which the old codes were applicable.

Twenty-five years ago, 3000 psi concrete was the standard in the construction industry. Today, concrete with 10,000 psi strength is used for lower story columns and walls of very tall high-rise buildings. Grade 40 reinforcing steel has been almost entirely replaced by Grade 60 reinforcement.

Gradual switching in the 1963 and 1971 Codes from the Working Stress Design Method to the Strength Design Method permitted more efficient designs of the structural components of buildings. The size of structural sections (columns, beams, and slabs) became substantially smaller and utilized less reinforcement, resulting in a 20 to 25% reduction in structural frame costs.

While we have seen dramatic increases in the strength of materials and greater cost efficiencies and design innovations made possible by the use of the strength design method, we have, as a consequence, also created new and more complex problems. The downsizing of structural components has reduced overall building stiffness. A further reduction has resulted from the replacement of heavy exterior wall cladding and partitions with lightweight materials which generally do not contribute to building stiffness. In particular, the drastic increase of stresses in the reinforcement at service loads from less than 20 ksi to more than 30 ksi caused a significantly wider spread of flexural cracking at service loads in slabs and beams, with consequent increase in their deflections.

When structures were designed by the classical working stress design, both strength and serviceability of the structure were ensured by limiting the stresses in the concrete and the reinforcement, in addition to imposing limits on slenderness ratios for the members. The introduction of strength design with the consequent more slender members significantly increased the design process; in addition to designing for strength, a separate consideration of serviceability (deflections and cracking) became necessary.

Complex structures require a more complex design approach. We are also now frequently dealing with bolder, larger, taller structures which are not only more complex, but also more flexible. Their structural behavior is characterized by larger deformations relative to member dimension than we have experienced before. As a consequence, a number of effects which heretofore were considered secondary and could be neglected, now became primary considerations during the design process. In this category are changes in geometry of structures due to gravity and lateral loadings. The effects of shrinkage, creep, and temperature are also becoming significant and can no longer be neglected in tall or in long structures, because of their cumulative effects.

1.4 A SIMPLE CODE

The more complex buildings undoubtedly require more complex design procedures to produce safe and economical structures. However, when we look at the reality of the construction industry as discussed at the beginning of this chapter, it makes little sense to impose on structures of moderate size (representing the bulk of construction volume) an intricate design approach which was developed to assure safety in highly complex structures. While the advances of the past decades have made it possible to build economical concrete structures reaching 1000 ft in height, the makeup of low-rise buildings has not changed significantly over the years.

It is possible to write a simplified code to be applicable to both moderate and large complex structures. However, this would require a technical conservatism in proportioning of members. While the cost of moderate structures would not be substantially affected by such an approach, the competitiveness of large complex structures could be severely impaired. To avoid such unnecessary circumstances, and at the same time to stay within required safety limits, it is possible to extract from the complex code a simplified design approach that can be applied to specifically defined moderate-sized structures. Such structures are characterized as having configurations and rigidity to eliminate sensitivity to secondary stresses and having members proportioned with sufficient conservatism to be able to simplify complex code provisions.

Second. . . .the Purpose and Scope of Simplified Design

1.5 PURPOSE OF SIMPLIFIED DESIGN

The purpose of this manual is to give practicing engineers some ways to reduce the design time required for smaller projects while still complying with the letter and intent of the ACI Standard 318-83 "Building Code Requirements for Reinforced Concrete." The simplification of design with its attendant savings in design time result from avoiding building member proportioning details and material property selections which make it necessary to consider certain complicated and complicating provisions of the ACI Standard. Often these situations can be avoided by making minor changes in design approach. In the various chapters of this book specific recommendations are made to accomplish this goal.

The simplified design procedures presented in this manual are an attempt to solve the various design considerations that need to be addressed in the structural design and detailing of primary framing members of a reinforced concrete building--by the simplest and quickest procedure possible. The simplified design material is intended for use by experienced engineers well-versed in the design principles of reinforced concrete and, completely familiar with the design provisions of the ACI 318 Standard for Reinforced Concrete. As noted in the foreword, this manual has been written solely as a design timesaver....to simplify design procedures using the provisions of the ACI 318 for reinforced concrete buildings of moderate size and height.

1.6 SCOPE OF SIMPLIFIED DESIGN

The simplified design approach presented in this manual should be used within the following general guidelines and limitations. In addition, appropriate guidelines and limitations are given within the Chapters for proper application of a specific simplifying design procedure:

- **Type of Construction:** Conventionally reinforced cast-in-place construction. Prestressed and precast construction are not addressed.

- **Building Size:** Buildings of moderate size and height with usual spans and story heights. Maximum building dimension should be in the range of 200 ft to 250 ft to reduce effects of shrinkage and temperature to manageable amounts. Maximum height of building should be in the range of 4 to 6 stories to justify the economics of simplified design.

- **Materials:** Normal weight concrete....f'_c = 4000 psi
  Deformed reinforcing bars....f_y = 60,000 psi

Both material strengths are readily available in the market place and will result in members that are durable and perform well structurally. One set of material parameters greatly simplifies the presentation of design aids. The 4000/60,000 strength combination is used in all simplified design expressions and design aids presented in this manual....with one exception; the simplified thickness design for footings addresses both f'_c = 3000 psi and f'_c = 4000 psi concrete strength.

In most cases, the designer can easily modify the simplified design expressions for other material strengths. Also, welded wire fabric and lightweight concrete may be used with the simplified design procedures, with appropriate modification as required by ACI 318.

• **Loadings:** Design dead load, live load, and wind forces are in accordance with American National Standards "Minimum Design Loads for Buildings and Other Structures" (ANSI A58.1-1982), with such reductions in live loads as permitted in ANSI A58.1. The local building code having jurisdiction in the locality of construction should be consulted for any possible differences in design loads from those stated in ANSI A58.1.

If resistance to earthquake-induced forces, earth or liquid pressure, impact effects, or structural effects of differential settlement, shrinkage, or temperature change need to be included in design, such effects are to be included separately, in addition to the normal loading of dead load, live load, and wind forces. (ACI 9.2.3 through 9.2.7)

Also, effects of forces due to snow loads, rain loads (ponding), and fixed service equipment (concentrated loads) are to be considered separately where applicable. (ACI 8.2.4)

Exposed exterior columns or open structures may require consideration of temperature change effects which are beyond the scope of this manual.

•**Design Method:** All simplified design procedures comply with provisions of "Building Code Requirements for Reinforced Concrete" (ACI 318-83), using appropriate load factors and strength reduction factors as specified in ACI 9.2 and 9.3. Reference to specific ACI Code provisions are noted....i.e. ACI 9.2, meaning ACI 318-83, Section 9.2.

Third. . . .the Application of Simplified Design

1.7 BUILDING EXAMPLES

To illustrate application of the simplified design approach presented in this manual, two building examples are included. **Building Example No. 1** is a 3-story one-way joist slab and column framing with two alternate joist floor systems; (1) standard pan joist and (2) spread (wide module) pan joist. **Building Example No. 2** is a 5-story two-way flat slab and column framing with two alternate wind-force resisting systems; (1) slab and column framing with spandrel beams and (2) structural walls.

To illustrate simplified design, typical structural members of the two buildings (beams, slabs, columns, walls, and footings) are designed by the simplified procedures presented in the various chapters of the manual. The design examples are located within each of the chapters where a simplified design condition is presented.

1.7.1 BUILDING EXAMPLE NO. 1 - 3-STORY PAN JOIST CONSTRUCTION

(1) Floor system: one-way joist slab

Alternate (1) - standard pan joists

Alternate (2) - spread (wide module) joists

(2) Wind-force resisting system: beam and column framing

(3) Load data: roof LL = 12 psf

DL = 105 psf (assume 95 psf joists and beams + 10 psf roofing and misc.)

floors LL = 60 psf

DL = 145 psf (assume 100 psf joists and beams + 20 psf partitions + 25 psf ceiling and misc.)

(4) Preliminary sizing:

Columns int. = 18x18

ext. = 16x16

Width of spandrel beams = 20 in.

Width of interior beams = 36 in.

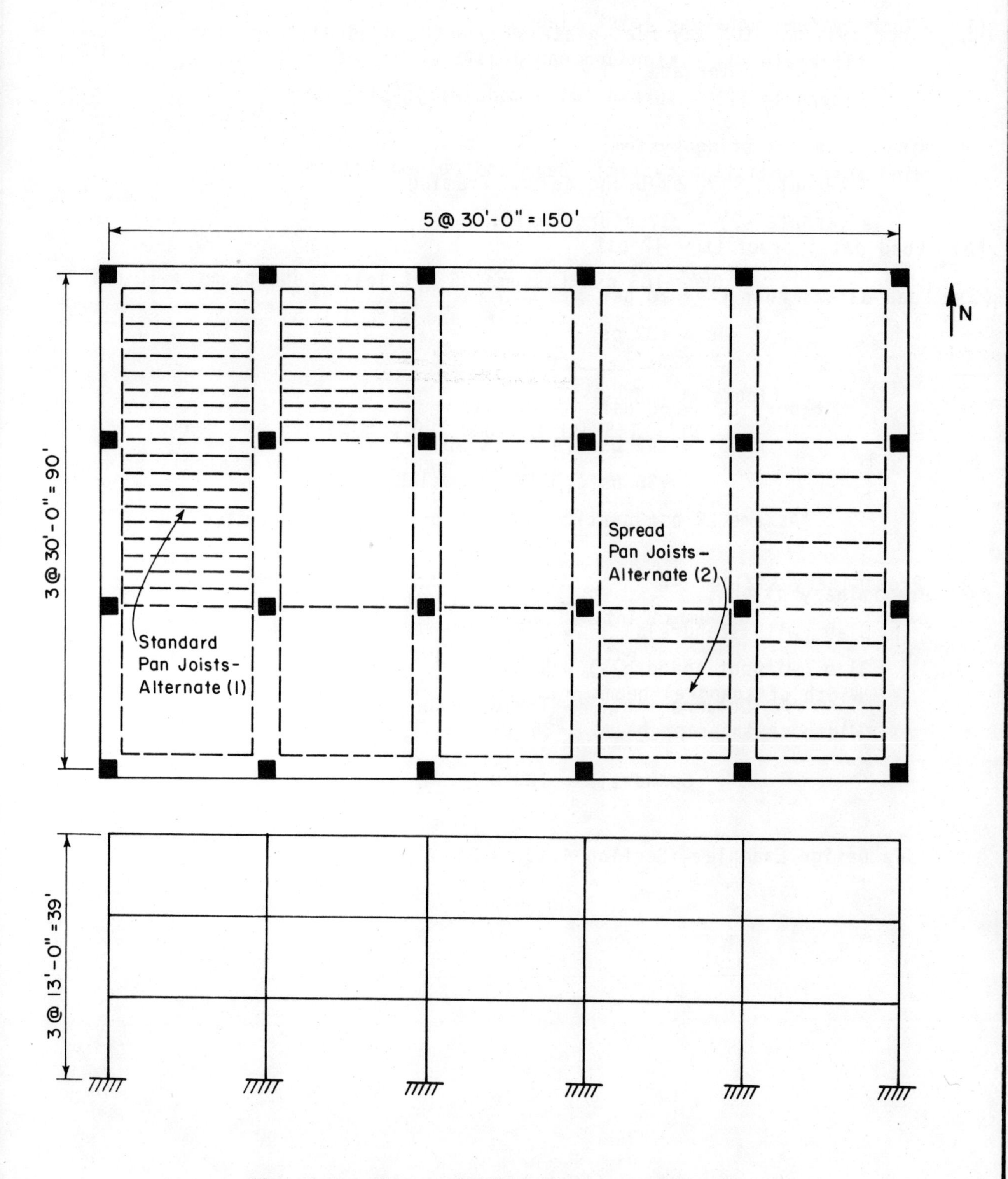

Fig. 1-3 Plan and Elevation for Bldg. #1

1.7.2 BUILDING EXAMPLE NO. 2 - 5-STORY FLAT PLATE CONSTRUCTION

(1) Floor system: two-way flat plate slab with spandrel beams for Alternate (1)

(2) Wind-force resisting system:

Alternate (1) - slab and column framing

Alternate (2) - structural walls

(3) Load data: roof LL = 20 psf

DL = 122 psf

floors LL = 50 psf

DL* = 142 psf (9 in. slab)

136 psf (8 1/2 in. slab)

*Assume 20 psf partitions + 10 psf ceiling and misc.

(4) Preliminary sizing:

Slab (with spandrels) = 8 1/2 in.

Slab (without spandrels) = 9 in.

Columns int. = 16x16

ext. = 12x12*

Spandrels = 12x20

*See Design Example - Section 4.7.1 - Step (8)

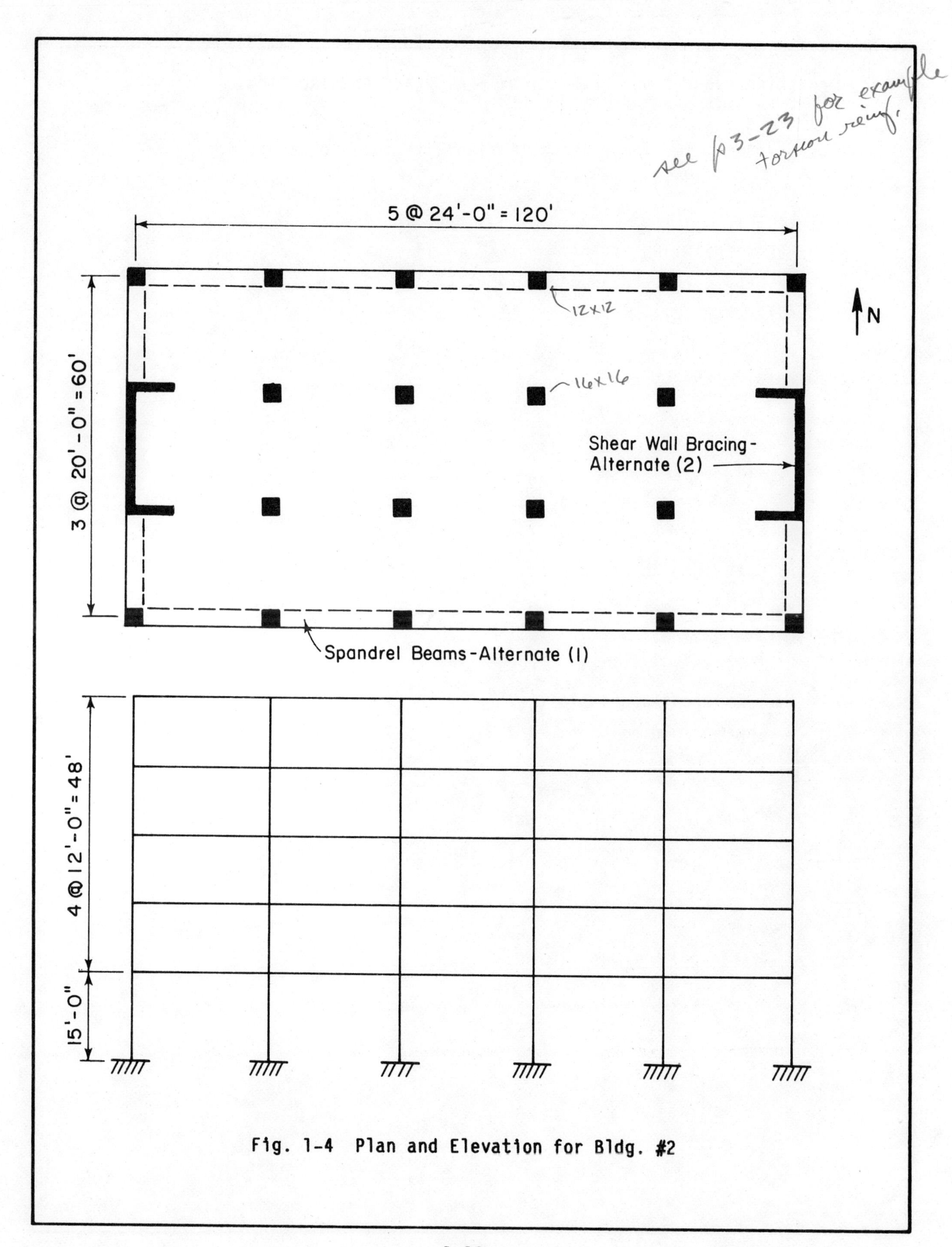

Fig. 1-4 Plan and Elevation for Bldg. #2

2

Simplified Frame Analysis

Randall C. Cronin*

2.1 INTRODUCTION

Final design of the structural components in the building frame is based on an analysis for maximum effects of moment, shear, axial load, and torsion or other load effects as generally determined by elastic frame analysis (ACI 8.3). For building frames of moderate size and height, preliminary and final designs often will be combined. Preliminary sizing of members, prior to analysis, may be based on designer experience, preliminary design aids, or simplified sizing expressions suggested in this manual.

Analysis of a structural frame or continuous construction is usually the most difficult part of the total design. For continuous one-way construction (beams and slabs), the approximate moments and shears (ACI 8.3.3) are satisfactory within the span and loading limitations stated. Also, for gravity load analysis of beam-column framing systems, a two-cycle moment distribution method is accurate enough. The speed and accuracy of the method can greatly simplify the gravity load analysis for building frames of usual types of construction, spans, and story heights. The method isolates one floor at a time and assumes the far ends of the upper and lower columns to be fixed. This simplifying assumption is permitted by ACI 8.8.3.

For lateral load analysis of an unbraced frame, the "Portal Method" offers a direct solution for the moments and shears in the beams (or slabs) and columns

*Senior Regional Structural Engineer, Eastern Region, PCA

due to wind loading without the need to consider size of members or member stiffnesses.

The simplified methods presented in this chapter for gravity load analysis and lateral wind load analysis of a structural frame or continuous construction are considered to provide sufficiently accurate results for buildings of moderate size and height. However, determination of load effects by computer analysis or other design aids are equally applicable for use with the simplified design procedures presented in subsequent chapters in this manual.

2.2 LOADING

2.2.1 Service Loads

The first step in the frame analysis is the determination of design (service) loads and wind forces as called for in the general building code under which the project is to be designed and constructed. For the purposes of this manual, design live loads (and permissible reductions in live loads) and wind loads are based on "Minimum Design Loads for Buildings and Other Structures," ANSI A58.1.[2.1] Reference to a specific ANSI code requirement is noted (i.e., ANSI 4.2, meaning ANSI A58.1-1982, Section 4.2). For a specific project, however, the governing general building code should be consulted for any difference from that of ANSI A58.1. Design for earthquake-induced load effects is not considered in this manual.

Design dead loads include member self-weight, weight of fixed service equipment (plumbing, electrical, etc.) and, where applicable, weight of built-in partitions. The latter may be accounted for by an equivalent uniform load of not less than 20 psf, although this is not specifically defined in the ANSI Code.

Design live loads will depend on the intended use and occupancy applicable to the portion or portions of the building being designed. Live loads include loads due to movable objects and movable partitions temporarily supported by the building during maintenance. In ANSI Table 2, uniformly distributed live loads range from 40 psf for residential use to 250 psf for heavy manufacturing and warehouse storage. Portions of buildings, such as library stacks and file rooms, require substantially heavier live loads. Live loads on a roof include maintenance equipment, workers, and materials. Also, snow loads, ponding of water, and special purposes, such as landscaping, must be included where applicable.

Occasionally, concentrated live loads must be included; however, they are more likely to affect individual supporting members and usually will not be included in the frame analysis.

Design wind loads are usually given in the general building code having jurisdiction. For building Example No. 1 and No. 2, calculation of wind loading is based on the procedure presented in ANSI 6.4.2.

2.2.1.1 - Example: Calculation of Wind Loads - Bldg. #2

For illustration of the ANSI procedure, which has been extensively revised with the '82 edition of the ANSI A58.1 Code, wind load calculations for the main wind-force resisting system for Bldg. #2 (5-story flatplate) are summarized as follows:

Wind-force resisting system:

Alternate (1) - Slab and column framing with spandrel beams

Alternate (2) - Structural walls

(1) Wind load data

Assume building located in Midwest in flat open terrain

Basic wind speed V = 80 mph	ANSI Fig. 1
Building exposure C = open terrain	ANSI 6.5.3
Importance factor I = 1.0	ANSI Table 1 & 5

Gust response factor G_h = 1.20 (h = 63 ft) ANSI Table 8

Wall pressure coefficients:

windward - both directions C_p = 0.8

leeward - EW-direction* C_p = 0.3 (L/B = 120/60 = 2)

- NS-direction* C_p = 0.5 (L/B = 60/120 = 0.5)

Maximum velocity pressure exposure coefficient K_h = 1.24 ANSI Table 6

*Due to building orientation in this example.

(2) Design wind pressures (NS-direction)

Height above ground level	K_z	q_z (psf)	Windward $q_zG_hC_p$	Leeward $q_hG_hC_p$	Total p_z (psf)
60-70	1.24	20.3	19.5	12.2	31.7
50-60	1.19	19.5	18.7	12.2	30.9
40-50	1.13	18.5	17.8	12.2	30.0
30-40	1.06	17.4	16.7	12.2	28.9
25-30	0.98	16.1	15.5	12.2	27.7
20-25	0.93	15.2	14.6	12.2	26.8
15-20	0.87	14.3	13.7	12.2	25.9
0-15	0.80	13.1	12.6	12.2	24.8

Sample calculations for 0-15 ft height range:

Design wind pressure p_z = (windward) + (leeward) ANSI Table 4

$p_z = (q_zG_hC_p) + (q_hG_hC_p)$

p_z = 12.6 + 12.2 = 24.8 psf

where $q_z = 0.00256\ K_z(IV)^2 = 0.00256 \times 0.80\ (80)^2 = 13.1$ psf

$q_zG_hC_p = 13.1 \times 1.20 \times 0.8 = \underline{12.6\ \text{psf}}$

and $q_h = 0.00256\ K_h(IV)^2 = 0.00256 \times 1.24\ (80)^2 = 20.3$ psf

$q_hG_hC_p = 20.3 \times 1.20 \times 0.5 = \underline{12.2\ \text{psf}}$ (leeward pressure uniform for full height of building)

(3) Lateral wind loads in NS-direction

Using the design wind pressure p_z, assumed uniform over the incremental heights above ground level, the following equivalent wind loads are calculated at each story level.

Alternate (1) Slab and column framing

Interior frame (24 ft bay width)

roof = 4.51^k
4th = 8.78^k
3rd = 8.44^k
2nd = 7.96^k
1st = 8.20^k

Alternate (2) Structural walls

Total in NS-direction (120 ft width)

roof = 22.6^k
4th = 43.9^k
3rd = 42.2^k
2nd = 39.8^k
1st = 41.0^k

(4) Lateral wind loads in EW-direction

Using the same procedure as for the NS-direction, the following story shears are obtained for the EW-direction

Alternate (1)	Alternate (2)
Interior frame (20 ft bay)	Total in EW-direction (60 ft width)
roof = 3.75^k	roof = 11.3^k
4th = 7.32^k	4th = 22.0^k
3rd = 7.04^k	3rd = 21.1^k
2nd = 6.64^k	2nd = 19.9^k
1st = 6.84^k	1st = 20.5^k

(5) The above wind load calculations assume a uniform design wind pressure p_z over the incremental heights above ground level as tabulated in ANSI Table 6, i.e., 0-15, 15-20, 20-25, etc. Alternatively, wind load calculations can be considerably simplified with results equally valid, especially for low-to-moderate height buildings, by computing design wind pressures at each floor level and assuming uniform pressure between midstory height above and below the floor level under consideration. For one- and two-story buildings, a design wind pressure computed at the roof height and assumed uniform for full building height would also seem accurate enough.

Recalculate the lateral wind loads in the NS-direction using design wind pressures computed at each floor level:

Story height above ground level (ft)	K_z	q_z (psf)	Windward $q_z G_h C_p$	Leeward $q_h G_h C_p$	Total p_z (psf)
roof - 63	1.21	19.8	19.0	11.9	30.9
4th - 51	1.14	18.7	18.0	11.9	29.9
3rd - 39	1.05	17.2	16.5	11.9	28.4
2nd - 27	0.94	15.4	14.8	11.9	26.7
1st - 15	0.80	13.1	12.6	11.9	24.5

where $q_h = 0.00256 K_h V^2 = 0.00256 \times 1.21 (80)^2 = 19.8$ psf

$q_h G_h C_p = 19.8 \times 1.20 \times 0.5 = 11.9$ psf

Lateral wind loads in NS-direction:

<u>Alternate (1) Slab and column framing</u>

Interior frame (24-ft width)

roof $= 30.9 \times 6 \times 24 = 4.45^k$

4th $= 29.9 \times 12 \times 24 = 8.61^k$

3rd $= 28.4 \times 12 \times 24 = 8.18^k$

2nd $= 26.7 \times 12 \times 24 = 7.69^k$

1st $= 24.5 \times 13.5 \times 24 = 7.94^k$

2.2.1.2 - Example: Calculation of Wind Loads - Bldg. #1

Wind load calculations for the main wind-force resisting system for Bldg. #1 (3-story pan joist) are summarized as follows:

Wind-force resisting system: Beam and column framing

(1) Wind load data

Assume building located in hurricane oceanline area in flat open terrain.

Basic wind speed V = 110 mph — ANSI Fig. 1

Building exposure C = open terrain — ANSI 6.5.3

Importance factor I = 1.11 — ANSI Table 1 & 5

(building Category II @ Hurricane oceanline)

Gust response factor G_h = 1.23 (h = 39'-0) — ANSI Table 8

Wall pressure coefficients:

Windward - both directions C_p = 0.8

Leeward - EW-direction* C_p = 0.3 (L/B = 150/90 = 1.67)

NS-direction* C_p = 0.5 (L/B = 90/150 = 0.6)

Maximum velocity pressure exposure coefficient K_h = 1.06 — ANSI Table 6

*Due to building orientation

(2) Summary of lateral wind loads

(a) NS & EW-direction (use same values for both directions)

Interior frame (30'-0 bay width)

Roof = 8.52^k

2nd = 16.02^k

1st = 14.29^k

2.2.2 Live Load Reduction for Columns, Beams, and Slabs

Most general building codes permit a reduction in live load for design of columns, beams and slabs to account for the probability that the total floor area "influencing" the load on a member may not be fully loaded simultaneously. Traditionally, the concept of reducing the amount of live load for which a member must be designed has been based on ~~arbitrary~~ tributary floor area supported by that member. With publication of the 1982 edition of ANSI A58.1, the concept of live load reduction is revised significantly to reflect an influence area rather than a tributary area for evaluating the amount of reduction permitted. For example, for an interior column the influence area is the total floor area of the four surrounding bays (four times the traditional tributary area). For an edge column, the two adjacent bays "influence" the load effects on the column and a corner column has an influence area of one bay. For interior beams the influence area consists of the two adjacent panels, while for the peripheral beam it is only one panel. For two-way slabs, the influence area is equal to the panel area.

The reduced live load L_r per square foot of floor area supported by columns, beams, and two-way slabs having an influence area of more than 400 sq ft is

$$L_r = L\left(0.25 + \frac{15}{\sqrt{A_I}}\right)$$

where L is the unreduced design live load per square foot, and A_I is the influence area as described above. The reduced live load cannot be taken less than 50% for members supporting one floor, nor less than 40% of the unit live load L otherwise. No reduction of live loads can be made for places of public assembly, for parking garages (except for garages limited to passenger cars only), one-way slabs, and for roofs. Influence areas for columns supporting more than one floor are summed.

Using the above expression for reduced live load, values of the reduction multiplier as a function of influence area are given in Table 2-1.

The live load reduction multiplier for beams and two-way slabs having an influence area of more than 400 sq ft ranges from 1.0 to 0.5. For influence areas on these members exceeding 3600 sq ft the reduction multiplier of 0.5 remains constant.

The live load reduction multiplier for columns in multistory buildings ranges from 1.0 to 0.4 for cumulative influence areas between 400 and 10,000 sq ft. For influence areas on columns exceeding 10,000 sq ft, the reduction multiplier of 0.4 remains constant.

The above discussion on permissible reduction in live loads is based on ANSI 4.7. The governing general building code should be consulted for any difference in amount of reduction and type of members that may be designed for a reduced live load.

Table 2-1 - Reduction Multiplier (RM) for Live Load ($0.25 + 15/\sqrt{A_I}$)

Influence Area A_I	RM	Influence Area A_I	RM
400[1]	1.000	5600	0.450
800	0.780	6000	0.444
1200	0.683	6400	0.438
1600	0.625	6800	0.432
2000	0.585	7200	0.427
2400	0.556	7600	0.422
2800	0.533	8000	0.418
3200	0.515	8400	0.414
3600	0.500[2]	8800	0.410
4000	0.487	9200	0.406
4400	0.476	9600	0.403
4800	0.467	10000	0.400[3]
5200	0.458		

[1]No live load reduction permitted for influence area less than 400 sq ft
[2]Maximum reduction permitted for members supporting one floor only
[3]Maximum absolute reduction

2.2.2.1 - Example: Live Load Reductions - Bldg. #2

For illustration, typical influence areas for the columns and end shear wall units of Bldg. #2 (5-story flatplate) are shown in Fig. 2-1. Corresponding live load reduction multipliers are tabulated in Table 2-2.

For example, the interior columns of the 1st story are designed for a reduced live load L_r = 0.42L (A_I = 4-bay areas x 4 stories = 20 x 24 x 4 x 4 = 7680 sq ft). The two-way slab may be designed with an RM = 0.94 (A_I = 480 for one-bay area). Shear strength around the interior columns is designed for an RM = 0.59 (A_I = 1920 for 4-bay areas), and around an edge column RM = 0.73 (A_I = 960 for 2-bay areas). Spandrel beams could be designed with an RM = 0.94 (one-bay area). If the floor system were a two-way slab with beams between columns, the interior beams would qualify for an RM = 0.73 (2-bay areas).

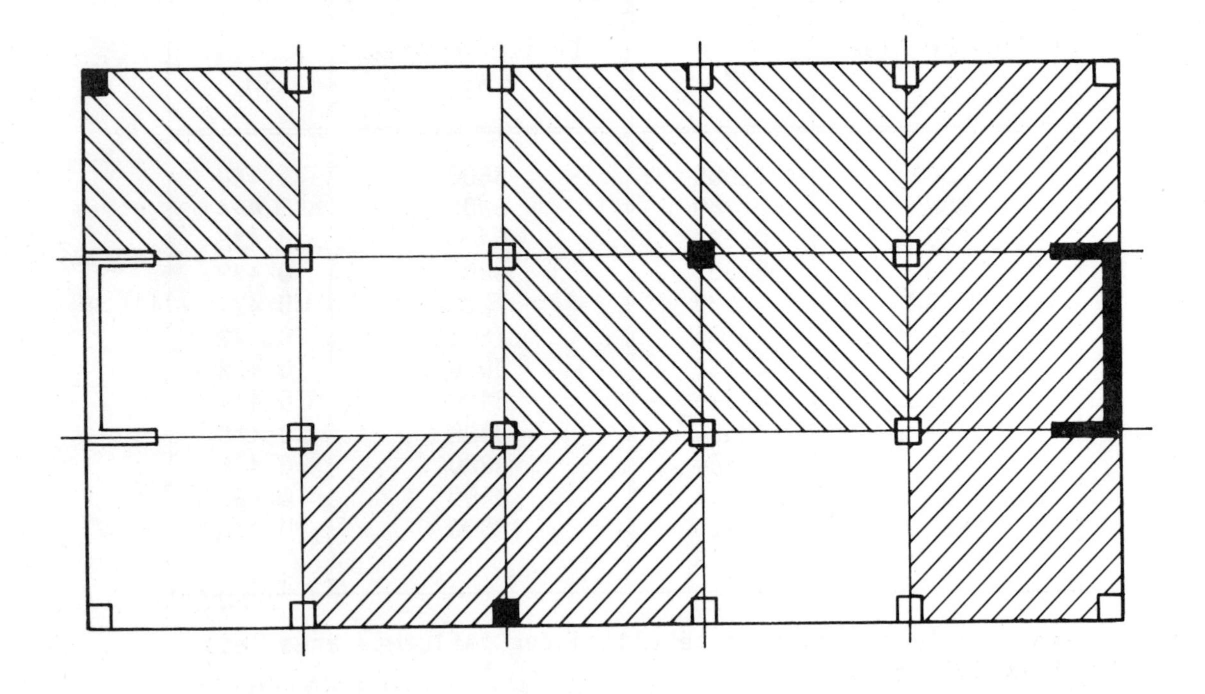

Fig. 2-1 Typical Influence Areas (Bldg. #2)

Table 2-2 - Reduction Multiplier for Live Loads (Bldg. #2)

Story	Interior Columns		Edge Columns		Corner Columns		End Shear Wall Units	
	A_I	RM	A_I	RM	A_I	RM	A_I	RM
5 (roof)		*		*		*		*
4	1920	0.59	960	0.73	480	0.94	1440	0.65
3	3840	0.49	1920	0.59	960	0.73	2880	0.53
2	5760	0.45	2880	0.53	1440	0.65	4320	0.48
1	7680	0.42	3840	0.49	1920	0.59	5760	0.45

*No reduction permitted for roof live load.

2.2.3 Factored Loads

The strength method of design, using factored loads to proportion members, is used exclusively in this manual. The design (service) loads must be increased by specified load factors (ACI 9.2) and factored loads combined in load combinations depending on the types of loads being considered.

For design of beams and slabs, the factored load combination used most often is:

$$U = 1.4D + 1.7L \qquad \text{ACI Eq. (9-1)}$$

$U_c = 1.6(D+L)$

For a frame analysis with live load applied only to a portion of the structure, i.e., alternate spans (ACI 8.9), the factored loads to be applied would be computed separately using the appropriate load factor for each load. However, for approximate methods of analysis, such as the approximate moment and shear expressions of ACI 8.3.3, where live load can be assumed to be applied over the entire structure, it may be expedient to use a composite load factor $U = C(D+L)$. An exact value of C can be computed for any combination of dead load to live load as:

$$C = \frac{1.4D + 1.7L}{D + L} = \frac{1.4 + 1.7(L/D)}{1 + L/D}$$

Table 2-3 gives composite load factors for various types of floor systems and building occupancies. If desired, a composite load factor may be taken directly from the table and interpolated, if necessary, for any usual L/D ratio. Alternately, a single value of C=1.6 could be used without much error from an exact value. An overdesign of about 9% for low L/D ratios and an underdesign of 1.6% for L/D=3 would result. Since the majority of building designs will not involve L/D ratios in the very low range or approaching the value of 3.0, use of the single value of C=1.6 for all designs can provide an effective simplification. For most structures, the moderate overdesign, resulting in slight increases of steel percentages, will enhance building performance.

Table 2-3 - Composite Load Factors

Floor system	D(psf)	Use	L(psf)	L/D	C		% diff.
9" Flat plate	120	roof	30	0.25	1.46	1.6	9.6
3' Waffle (25' span)	120	resid.	40	0.33	1.47		8.8
5' Waffle (50' span)	175	office	70	0.40	1.49		7.4
5" Flat plate	70	roof	35	0.50	1.50		6.7
5" Flat plate	60	resid.	40	0.67	1.52		5.3
5' Waffle (25' span)	100	office	100	1.00	1.55		3.2
5' Waffle (30' span)	173	indust.	250	1.45	1.58		1.3
3' Waffle (20' span)	125	indust.	250	2.00	1.60		-
Joists (15' x 25')	100	indust.	250	2.50	1.61		-0.6
5' Waffle (25' span)	133	library	400	3.00	1.625	1.6	-1.6

There is one final consideration when using factored loads to proportion members. The designer has the choice of multiplying the service loads by the load factors before computing the factored load effects (moments, shears, etc.), or computing the effects from the service loads and multiplying the effects by the load factors. For example, in the computation of bending moment for dead and live load [U = 1.4D + 1.7L or U = 1.6 (D+L)], the designer may (1) determine $w_u = 1.4\ w_d + 1.7\ w_\ell$ and then compute the factored moments using w_u; or (2) compute the dead and live load moments using service loads and then determine the factored moments as $M_u = 1.4M_d + 1.7M_\ell$. Both analysis procedures yield the same answer.

2.3 FRAME ANALYSIS BY COEFFICIENTS

The ACI Code provides a simplified method of analysis for both one-way construction (ACI 8.3.3) and two-way construction (ACI 13.6). Both simplified methods provide for determination of moments and shears based on coefficients. Each method will give satisfactory results within the span and loading conditions stated. The direct design method for two-way slabs is discussed in Chapter 4.

2.3.1 Continuous Beams and One-Way Slabs

When beams and one-way slabs are part of a frame or continuous construction, ACI 8.3.3 provides approximate moment and shear coefficients for gravity load analysis. The approximate coefficients apply within the conditions illustrated in Fig. 2-2. There must be two or more spans, approximately equal in length, with the longer of two adjacent spans not exceeding the shorter by more than 20 percent; loads must be uniformly distributed, with the live load not more than 3 times the dead (L/D ≤ 3); and, members must have uniform cross section throughout the span. Also, no redistribution of moments (ACI 8.4) is permitted. The moment coefficients as defined in ACI 8.3.3 are shown in Figs. 2-3 through 2-7. In all cases, the shear in end span members at the interior support is taken equal to .... $1.15\ w_u\ell_n/2$. The shear at all other supports is .... $w_u\ell_n/2$ (see Fig. 2-7). The load term w_u is the factored load combination for dead and live load, $w_u = 1.4w_d + 1.7w_\ell$. For beams, w_u is the uniformly distributed load (plf), and the coefficients yield total moments and shears on the beam. For one-way slabs, w_u is the uniformly distributed load (psf), and the moments and shears are for slab strips one ft in width. The span length ℓ_n is always the clear span of the beam or slab. For negative moment at a support with unequal adjacent spans, ℓ_n is the average of the adjacent spans. At supports, moments and shears are at the face of support.

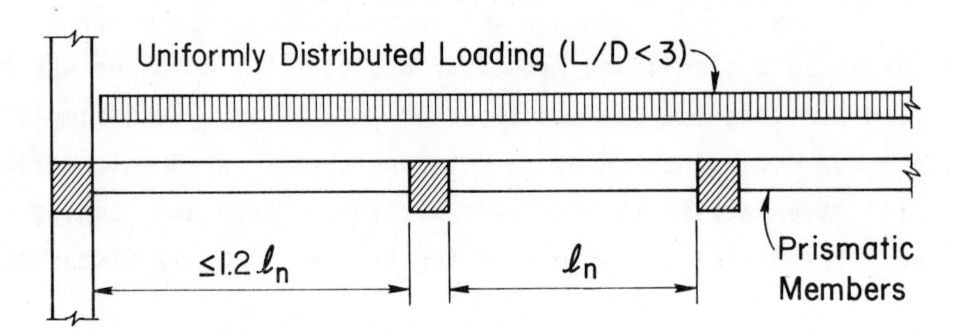

Fig. 2-2 Conditions for Analysis by Coefficients

2.3.1.1 - Example: Frame Analysis by Coefficients

Determine factored moments and shears for the joists of the standard pan joist floor system of Bldg. #1 (3-story pan joist) using approximate moment and shear coefficients of ACI 8.3.3.

(1) Data: Width of exterior (spandrel) beam = ~~18~~ 20 in.

Width of interior beams = 36 in.

Floors: LL = 60 psf

DL = 145 psf

w_u = 1.4 (145) + 1.7 (60) = 305 psf x 3 = 915 plf

(2) Factored shears and moments using coefficients from Figs. 2-3, 2-4, and 2-7 are computed as follows:

30'-0" | 30'-0" | 15'-0"

8" | 1'-0" | ℓ_n = 27.5' | 3'-0" | ℓ_n = 27.0' | 3'-0" | ℓ_n = 27.0'

Sym. about ℄

Total load	$w_u\ell_n$ = 0.92^k x 27.5 = 25.3^k		0.92 x 27.0 = 24.8^k		0.92 x 27.0 = 24.8^k
Coeff. shear V_u	1/2 12.7^k	1.15/2 14.5^k	1/2 12.4^k	1/2 12.4^k	1/2 12.4^k
$w_u\ell_n^2$ Coeff. Pos. M_u	695.8$'^k$ 1/14 49.7$'^k$		670.7$'^k$ 1/16 41.9$'^k$		670.7$'^k$ 1/16 41.9$'^k$
ℓ_n $w_u\ell_n^2$ Coeff. Neg. M_u	27.5 695.8^k 1/24 29.0$'^k$	27.25 683.2^k 1/10 68.3$'^k$	27.25 683.2^k 1/11 62.1$'^k$	27.0 670.7^k 1/11 61.0$'^k$	27.0 670.7^k 1/11 61.0$'^k$

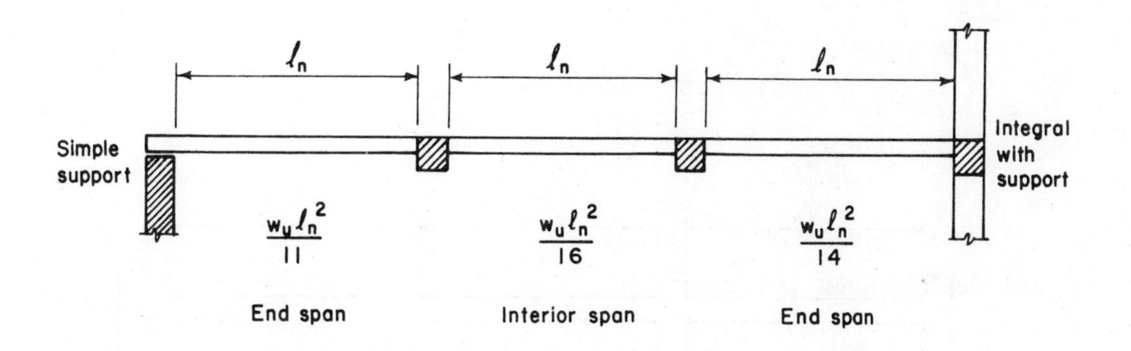

Fig. 2-3 - Positive Moments - All Cases

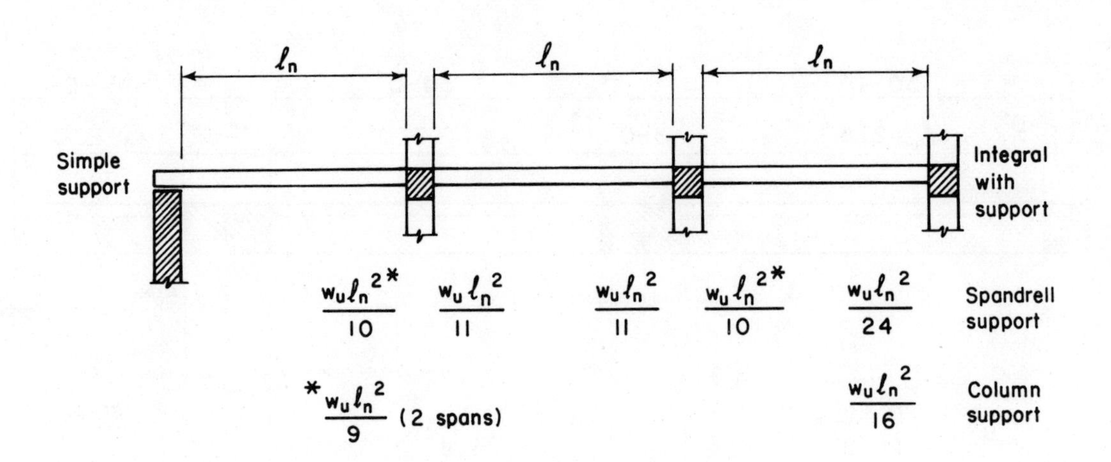

Fig. 2-4 - Negative Moments - Beams and Slabs

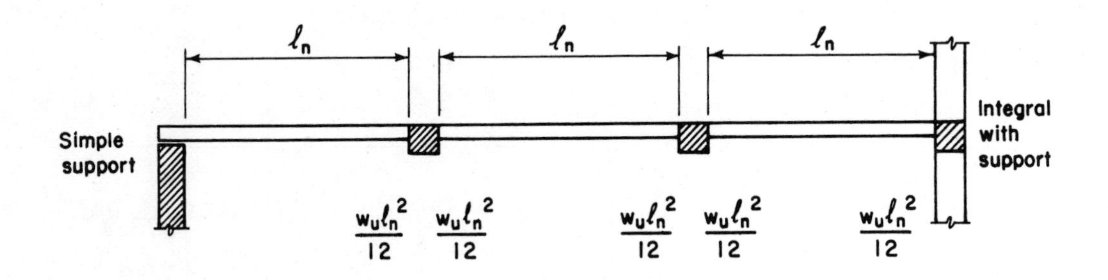

Fig. 2-5 - Negative Moments - Slabs with spans $\leq$ 10 ft

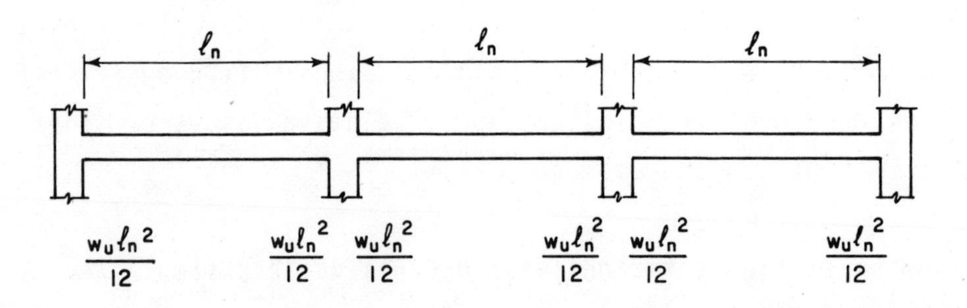

Fig. 2-6 - Negative Moments - Beams with Stiff Columns ($\Sigma K_c/\Sigma K_b$ > 8)

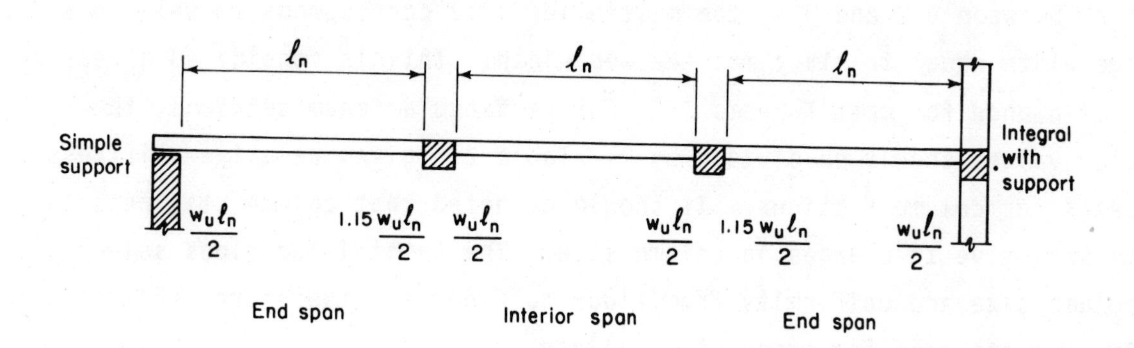

Fig. 2-7 End Shears - All Cases

2.4 FRAME ANALYSIS BY ANALYTICAL METHODS

For continuous beams and one-way slabs not meeting the conditions of ACI 8.3.3 for analysis by coefficients, an elastic frame analysis must be used. Approximate methods of frame analysis are permitted by ACI 8.3.2 for "usual" types of buildings. Simplifying assumptions on member stiffnesses, span lengths, and arrangement of live load are given in ACI 8.6 through 8.9.

2.4.1 Stiffness

The relative stiffness of frame members must be established regardless of the analytical method used. Any reasonable consistent procedure for determining stiffness of columns, walls, beams, and slabs is permitted by ACI 8.6. For buildings of moderate size and height, only relative values of stiffness will be used, for both braced and unbraced frames.

The selection of stiffness factors will be considerably simplified by use of Tables 2-4 and 2-5. The stiffness factors are based on gross section properties (neglecting any reinforcement) and should yield satisfactory results for buildings within the size and height range addressed by this manual. In most cases where an analytical analysis procedure is required, stiffness of T-beam sections will be required (i.e., transverse beams in a one-way joist floor). The relative stiffness values K given in Table 2-4 allow for the effect of the flange by doubling the moment of inertia of the web section $b_w h$. For values of h_f/h between 0.2 and 0.4, the multiplier of 2 corresponds closely to a flange width equal to six times the web width. This is considered a reasonable allowance for most T-beams.[2.2] For rectangular beam sections, the tabular values should be divided by 2. Table 2-5 gives relative stiffness values K for column sections. It should be noted that column stiffness is quite sensitive to changes in column size. The initial judicious selection of column size and uniformity from floor to floor is, therefore, critical in minimizing the need for successive analyses.

As is customary for ordinary building frames, torsional stiffness of transverse beams is not considered in the analysis. For those unusual cases where equilibrium torsion is involved, a more exact procedure may be necessary.

2.4.2 Arrangement of Live Load

Consider live load to be applied only to the floor or roof under consideration, with the ends of columns assumed fixed (ACI 8.9.1). In the usual case where the exact loading pattern is not known, develop maximum moments with loading conditions illustrated by the three-span partial frame in Fig. 2-8 and described as follows:

(a) When the service live load does not exceed three-quarters of the service dead load ($L/D \leq 3/4$), consider only loading pattern (1) with full live load on all spans for maximum positive and negative moments.

(b) When the service live-to-dead load ratio exceeds three-quarters, loadings patterns (2) through (4) need to be considered to determine all maximum moments.

Note: Live load reduction may apply to beams, see Section 2.2.2.

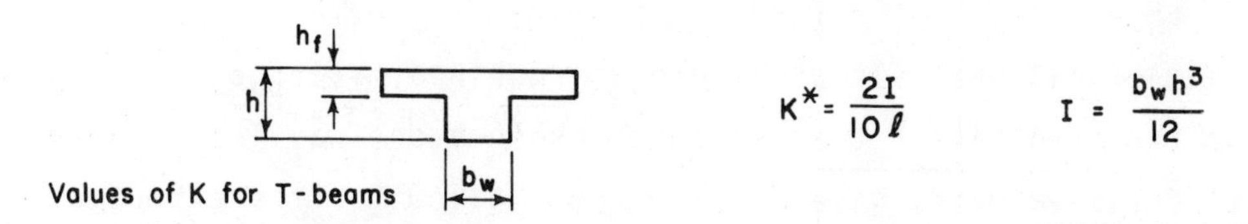

Values of K for T-beams

h	b_w	I	ℓ : Length of beam (feet)							
			8	10	12	14	16	20	24	30
8	6	256	6	5	4	4	3	3	2	2
	8	341	9	7	6	5	4	3	3	2
	10	427	11	9	7	6	5	4	4	3
	11½	491	12	10	8	7	6	5	4	3
	13	555	14	11	9	8	7	6	5	4
	15	640	16	13	11	9	8	6	5	4
	17	725	18	15	12	10	9	7	6	5
	19	811	20	16	14	12	10	8	7	5
10	6	500	13	10	8	7	6	5	4	3
	8	667	17	13	11	10	8	7	6	4
	10	833	21	17	14	12	10	8	7	6
	11½	958	24	19	16	14	12	10	8	6
	13	1083	27	22	18	15	14	11	9	7
	15	1250	31	25	21	18	16	13	10	8
	17	1417	35	28	24	20	18	14	12	9
	19	1583	40	32	26	23	20	16	13	11
12	6	864	22	17	14	12	11	9	7	6
	8	1152	29	23	19	16	14	12	10	8
	10	1440	36	29	24	21	18	14	12	10
	11½	1656	41	33	28	24	21	17	14	11
	13	1872	47	37	31	27	23	19	16	12
	15	2160	54	43	36	31	27	22	18	14
	17	2448	61	49	41	35	31	25	20	16
	19	2736	68	55	46	39	34	27	23	18
14	6	1372	34	27	23	20	17	14	11	9
	8	1829	46	37	30	26	23	18	15	12
	10	2287	57	46	38	33	29	23	19	15
	11½	2630	66	53	44	38	33	26	22	18
	13	2973	74	59	50	42	37	30	25	20
	15	3430	86	69	57	49	43	34	29	23
	17	3887	97	78	65	56	49	39	32	26
	19	4345	109	87	72	62	54	43	36	29
16	6	2048	51	41	34	29	26	20	17	14
	8	2731	68	55	46	39	34	27	23	18
	10	3413	85	68	57	49	43	34	28	23
	11½	3925	98	79	65	56	49	39	33	26
	13	4437	111	89	74	63	55	44	37	30
	15	5120	128	102	85	73	64	51	43	34
	17	5803	145	116	97	83	73	58	48	39
	19	6485	162	130	108	93	81	65	54	43
18	6	2916	73	58	49	42	36	29	24	19
	8	3888	97	78	65	56	49	39	32	26
	10	4860	122	97	81	69	61	49	41	32
	11½	5589	140	112	93	80	70	56	47	37
	13	6318	158	126	105	90	79	63	53	42
	15	7290	182	146	122	104	91	73	61	49
	17	8262	207	165	138	118	103	83	69	55
	19	9234	231	185	154	132	115	92	77	62
20	6	4000	100	80	67	57	50	40	33	27
	8	5333	133	107	89	76	67	53	44	36
	10	6667	167	133	111	95	83	67	56	44
	11½	7667	192	153	128	110	96	77	64	51
	13	8667	217	173	144	124	108	87	72	58
	15	10000	250	200	167	143	125	100	83	67
	17	11333	283	227	189	162	142	113	94	76
	19	12667	317	253	211	181	158	127	106	84
22	6	5324	133	106	89	76	67	53	44	36
	8	7099	177	142	118	101	89	71	59	47
	10	8873	222	177	148	127	111	89	74	59
	11½	10204	255	204	170	146	128	102	85	68
	13	11535	288	231	192	165	144	115	96	77
	15	13310	333	266	222	190	166	133	111	89
	17	15085	377	302	251	215	189	151	126	101
	19	16859	421	337	281	241	211	169	141	112

h	b_w	I	ℓ : Length of beam (feet)							
			8	10	12	14	16	20	24	30
24	8	9216	230	185	155	130	115	90	75	60
	10	11520	290	230	190	165	145	115	95	75
	11½	13248	330	265	220	190	165	130	110	90
	13	14976	375	300	250	215	185	150	125	100
	15	17280	430	345	290	245	215	175	145	115
	17	19584	490	390	325	280	245	195	165	130
	19	21888	545	440	365	315	275	220	180	145
	21	24192	605	485	405	345	300	240	200	160
26	8	11717	295	235	195	165	145	115	100	80
	10	14647	365	295	245	210	185	145	120	100
	11½	16844	420	335	280	240	210	170	140	110
	13	19041	475	380	315	270	240	190	160	125
	15	21970	550	440	365	315	275	220	185	145
	17	24899	620	500	415	355	310	250	205	165
	19	27829	695	555	465	400	350	280	230	185
	21	30758	770	615	515	440	385	310	255	205
28	8	14635	365	295	245	210	185	145	120	100
	10	18293	455	365	305	260	230	185	150	120
	11½	21037	525	420	350	300	265	210	175	140
	13	23781	595	475	395	340	295	240	200	160
	15	27440	685	550	455	390	345	275	230	185
	17	31099	775	620	520	445	390	310	260	205
	19	34757	870	695	580	495	435	350	290	230
	21	38416	960	770	640	550	480	385	320	255
30	8	18000	450	360	300	255	225	180	150	120
	10	22500	565	450	375	320	280	225	190	150
	11½	25875	645	520	430	370	325	260	215	175
	13	29250	730	585	490	420	365	295	245	195
	15	33750	845	675	565	480	420	340	280	225
	17	38250	955	765	640	545	480	385	320	255
	19	42750	1070	855	715	610	535	430	355	285
	21	47250	1180	945	790	675	590	475	395	315
36	8	31104	780	620	520	445	390	310	260	205
	10	38880	970	780	650	555	485	390	325	260
	11½	44712	1120	895	745	640	560	445	375	300
	13	50544	1260	1010	840	720	630	505	420	335
	15	58320	1460	1170	970	835	730	585	485	390
	17	66096	1650	1320	1100	945	825	660	550	440
	19	73872	1850	1480	1230	1060	925	740	615	490
	21	81648	2040	1630	1360	1170	1020	815	680	545
42	8	49392	1230	990	825	705	615	495	410	330
	10	61740	1540	1230	1030	880	770	615	515	410
	11½	71001	1780	1420	1180	1010	890	710	590	475
	13	80262	2010	1610	1340	1150	1000	805	670	535
	15	92610	2320	1850	1540	1320	1160	925	770	615
	17	104958	2620	2100	1750	1500	1310	1050	875	700
	19	117306	2930	2350	1950	1680	1470	1170	975	780
	21	129654	3240	2590	2160	1850	1620	1300	1080	865
48	8	73728	1840	1470	1230	1050	920	735	615	490
	10	92160	2300	1840	1540	1320	1150	920	770	615
	11½	105984	2650	2120	1770	1510	1320	1060	885	705
	13	119808	3000	2400	2000	1710	1500	1200	1000	800
	15	138240	3460	2760	2300	1970	1730	1380	1150	920
	17	156672	3920	3130	2610	2240	1960	1570	1310	1040
	19	175104	4380	3500	2920	2500	2190	1750	1460	1170
	21	193536	4840	3870	3230	2760	2420	1940	1610	1290
54	8	104976	2620	2100	1750	1500	1310	1050	875	700
	10	131220	3280	2620	2190	1880	1640	1310	1090	875
	11½	150903	3770	3020	2510	2160	1890	1510	1260	1010
	13	170586	4260	3410	2840	2440	2130	1710	1420	1140
	15	196830	4920	3940	3280	2810	2460	1970	1640	1310
	17	223074	5580	4460	3720	3190	2790	2230	1860	1490
	19	249318	6230	4990	4160	3560	3120	2490	2080	1660
	21	275562	6890	5510	4590	3940	3440	2760	2300	1840

*Coefficient 10 introduced to reduce magnitude of relative stiffness values

Table 2-4 - Beam Stiffness Factors

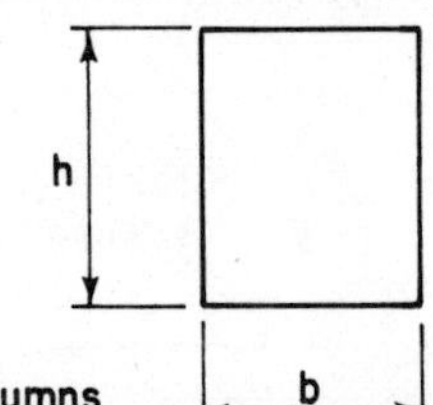

$$K^* = \frac{I}{10\,\ell_c} \qquad I = \frac{bh^3}{12}$$

Values of K for columns

h	b	I	ℓ_c: Height of column (feet) 8	9	10	11	12	14	16	20
8	10	427	5	5	4	4	4	3	3	2
	12	512	6	6	5	5	4	4	3	3
	14	597	7	7	6	5	5	4	4	3
	18	766	10	9	8	7	6	5	5	4
	22	939	12	10	9	9	8	7	6	5
	26	1109	14	12	11	10	9	8	7	6
	30	1280	16	14	13	12	11	9	8	6
	36	1536	19	17	15	14	13	11	10	8
10	10	833	10	9	8	8	7	6	5	4
	12	1000	13	11	10	9	8	7	6	5
	14	1167	15	13	12	11	10	8	7	6
	18	1500	19	17	15	14	13	11	9	8
	22	1833	23	20	18	17	15	13	11	9
	26	2167	27	24	22	20	18	16	14	11
	30	2500	31	28	25	23	21	18	16	13
	36	3000	38	33	30	27	25	21	19	15
12	10	1440	18	16	14	13	12	10	9	7
	12	1728	22	19	17	16	14	12	11	9
	14	2016	25	22	20	18	17	14	13	10
	18	2592	32	29	26	24	22	19	16	13
	22	3168	40	35	32	29	26	23	20	16
	26	3744	47	42	37	34	31	27	23	19
	30	4320	54	48	43	39	36	31	27	22
	36	5184	65	58	52	47	43	37	32	26
14	10	2287	29	25	23	21	19	16	14	11
	12	2744	34	30	27	25	23	20	17	14
	14	3201	40	36	32	29	27	23	20	16
	18	4116	51	46	41	37	34	29	26	21
	22	5031	63	56	50	46	42	36	31	25
	26	5945	74	66	59	54	50	42	37	30
	30	6860	86	76	69	62	57	49	43	34
	36	8232	103	91	82	75	69	59	51	41
16	10	3413	43	38	34	31	28	24	21	17
	12	4096	51	46	41	37	34	29	26	20
	14	4779	60	53	48	43	40	34	30	24
	18	6144	77	68	61	56	51	44	38	31
	22	7509	94	83	75	68	63	54	47	38
	26	8875	111	99	89	81	74	63	55	44
	30	10240	128	114	102	93	85	73	64	51
	36	12288	154	137	123	112	102	88	77	61
18	10	4860	61	54	49	44	41	35	30	24
	12	5832	73	65	58	53	49	42	36	29
	14	6804	85	76	68	62	57	49	43	34
	18	8748	109	97	87	80	73	62	55	44
	22	10692	134	119	107	97	89	76	67	53
	26	12636	158	140	126	115	105	90	79	63
	30	14580	182	162	146	133	122	104	91	73
	36	17496	219	194	175	159	146	125	109	87
20	10	6667	83	74	67	61	56	48	42	33
	12	8000	100	89	80	73	67	57	50	40
	14	9333	117	104	93	85	78	67	58	47
	18	12000	150	133	120	109	100	86	75	60
	22	14667	183	163	147	133	122	105	92	73
	26	17333	217	193	173	158	144	124	108	87
	30	20000	250	222	200	182	167	143	125	100
	36	24000	300	267	240	218	200	171	150	120
22	10	8873	111	99	89	81	74	63	55	44
	12	10648	133	118	106	97	89	76	67	53
	14	12422	155	138	124	113	104	89	78	62
	18	15972	200	177	160	145	133	114	100	80
	22	19521	244	217	195	177	163	139	122	98
	26	23071	288	256	231	210	192	165	144	115
	30	26620	333	296	266	242	222	190	166	133
	36	31944	399	355	319	290	266	228	200	160

h	b	I	ℓ_c: Height of column (feet) 8	9	10	11	12	14	16	20
24	12	13824	175	155	140	125	115	100	85	70
	14	16128	200	180	160	145	135	115	100	80
	18	20738	260	230	205	190	175	150	130	105
	22	25344	315	280	255	230	210	180	160	125
	26	29952	375	335	300	270	250	215	185	150
	30	34560	430	385	345	315	290	245	215	175
	36	41472	520	460	415	375	345	295	260	205
	42	48384	605	540	485	440	405	345	300	240
26	12	17576	220	195	175	160	145	125	110	90
	14	20505	255	230	205	185	170	145	130	105
	18	26364	330	295	265	240	220	190	165	130
	22	32223	405	360	320	295	270	230	200	160
	26	38081	475	425	380	345	315	270	240	190
	30	43940	550	490	440	400	365	315	275	220
	36	52728	660	585	525	480	440	375	330	265
	42	61516	770	685	615	560	515	440	385	310
28	12	21952	275	245	220	200	185	155	135	110
	14	25611	320	285	255	235	215	185	160	130
	18	32928	410	365	330	300	275	235	205	165
	22	40245	505	445	400	365	335	285	250	200
	26	47563	595	530	475	430	395	340	295	240
	30	54880	685	610	550	500	455	390	345	275
	36	65856	825	730	660	600	550	470	410	330
	42	76832	960	855	770	700	640	550	480	385
30	12	27000	340	300	270	245	225	195	170	135
	14	31500	395	350	315	285	265	225	195	160
	18	40500	505	450	405	370	340	290	255	205
	22	49500	620	550	495	450	415	355	310	250
	26	58500	730	650	585	530	490	420	365	295
	30	67500	845	750	675	615	565	480	420	340
	36	81000	1010	900	810	735	675	580	505	405
	42	94500	1180	1050	945	860	790	675	590	475
32	12	32768	410	365	330	300	275	235	205	165
	14	38229	480	425	380	350	320	275	240	190
	18	49152	615	545	490	445	410	350	305	245
	22	60075	750	670	600	545	500	430	375	300
	26	70997	885	790	710	645	590	505	445	355
	30	81920	1020	910	820	745	685	585	510	410
	36	98304	1230	1090	985	895	820	700	615	490
	42	114688	1430	1270	1150	1040	955	820	715	575
34	12	39304	490	435	395	355	330	280	245	195
	14	45855	575	510	460	415	380	330	285	230
	18	58956	735	655	590	535	490	420	370	295
	22	72057	900	800	720	655	600	515	450	360
	26	85159	1060	945	850	775	710	610	530	425
	30	98260	1230	1090	985	895	820	700	615	490
	36	117912	1470	1310	1180	1070	980	840	735	590
	42	137564	1720	1530	1380	1250	1150	985	860	690
36	12	46656	585	520	465	425	390	335	290	235
	14	54432	680	605	545	495	455	390	340	270
	18	69984	875	780	700	635	585	500	435	350
	22	85536	1070	950	855	780	715	610	535	430
	26	101088	1260	1120	1010	920	840	720	630	505
	30	116640	1460	1300	1170	1060	970	835	730	585
	36	139968	1750	1560	1400	1270	1170	1000	875	700
	42	163296	2040	1810	1630	1480	1360	1170	1020	815
38	12	54872	685	610	550	500	460	390	345	275
	14	64017	800	710	640	580	535	455	400	320
	18	82308	1030	915	825	750	685	590	515	410
	22	100599	1260	1120	1010	915	840	720	630	505
	26	118889	1490	1320	1190	1080	990	850	745	595
	30	137180	1710	1520	1370	1250	1140	980	855	685
	36	164616	2060	1830	1650	1500	1370	1180	1030	825
	42	192052	2400	2130	1920	1750	1600	1370	1200	960

*Coefficient 10 introduced to reduce magnitude of relative stiffness values

Table 2-5 - Column Stiffness Factors

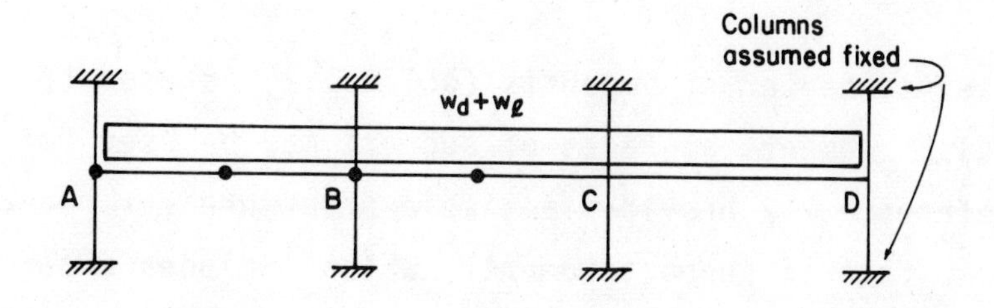

(1) Loading pattern for moments in all spans (L/D ≤ 3/4)

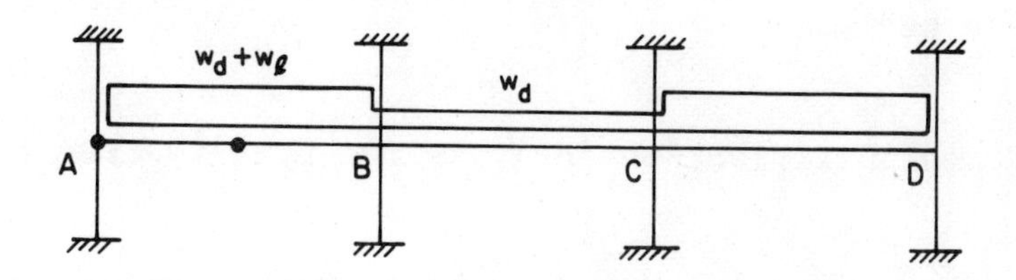

(2) Loading pattern for negative moment at support A and positive moment in span AB

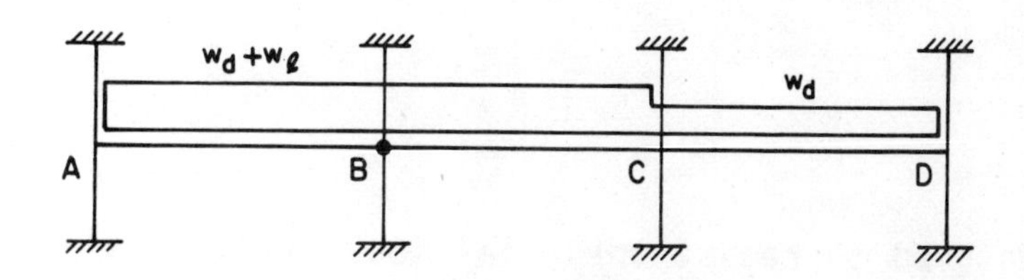

(3) Loading pattern for negative moment at support B

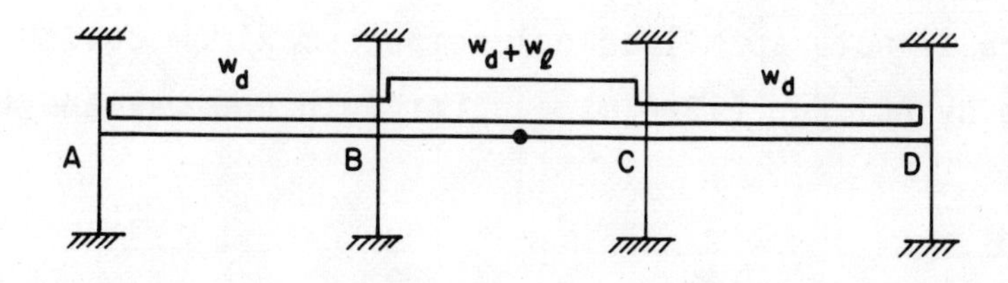

(4) Loading pattern for positive moment in span BC

Fig. 2-8 Partial Frame Analysis for Gravity Loading

2.4.3 Design Moments

Span length center-to-center of supports (ACI 8.7.2) is used for members in the frame analysis. Moments at faces of support may be used for member design (ACI 8.7.3). Reference 2.2 provides a simple procedure for reducing the centerline moments to face of support moments, which includes a correction for the increased end moments in the beam due to the restraining effect of the column between face and centerline of support. Fig. 2-9 illustrates this correction. For beam and slab design, negative moments from the frame analysis can be reduced by $w_u \ell^2 a/6$. A companion reduction in the positive moment of $w_u \ell^2 a/12$ can also be made.

2.4.4 Two-Cycle Moment Distribution Analysis for Gravity Loading

Reference 2.2 presents a simplified two-cycle method of moment distribution for ordinary building frames. The method meets the requirements for an elastic analysis called for in ACI 8.3 with the simplifying assumptions of ACI 8.6 through 8.9. The speed and accuracy of the two-cycle method will be of great assistance to designers. For an in-depth discussion of the principles involved, the reader is referred to Reference 2.2.

2.5 COLUMNS

Columns must be designed to resist the axial loads from gravity loading on all floors and roof above plus the maximum moment from gravity loads on a single adjacent span of the floor or roof under consideration.

For interior columns (unless a general analysis is made to evaluate gravity load moments from alternate span loading) compute maximum column moments due to gravity loading by ACI Eq. (13-4).......for both one-way and two-way floor systems:

$$M_u = 0.07\ [w_d(\ell_n^2 - \ell_n'^2) + 0.5\ w_\ell \ell_n^2]$$

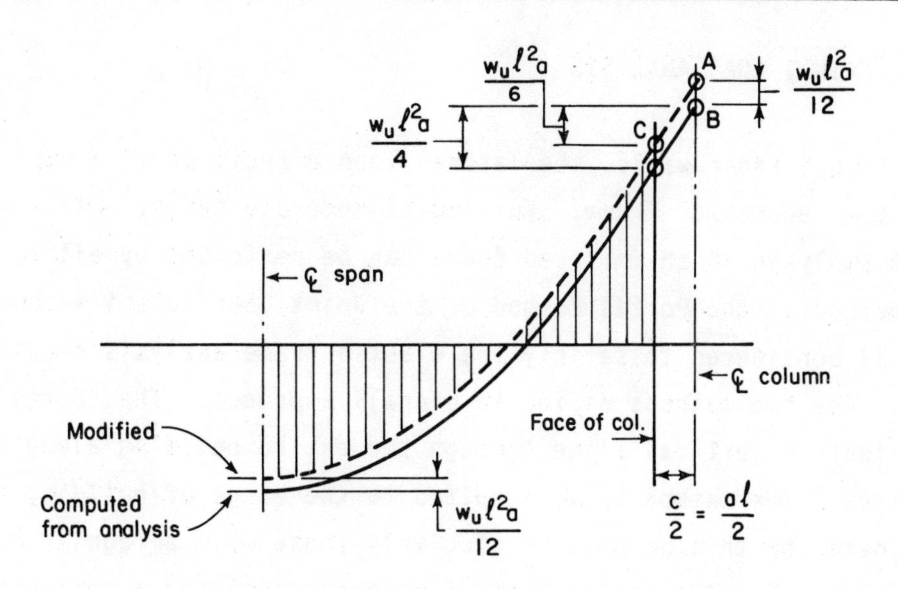

(A) = Theoretical ℄ moment
(B) = Computed ℄ moment ignoring stiffening effect of column support
(C) = Modified moment at face of column

w_u = uniformly distributed factored load (plf)
ℓ = span length center-to-center of supports
c = width of column support
a = c/ℓ

Fig. 2-9 Correction Factors for Span Moments[2.2]

where, for one-way construction:

w_d = uniformly distributed factored dead load, plf
w_ℓ = uniformly distributed factored live load (including any live load reduction; see Section 2.2.2), plf
ℓ_n = clear span length of longer adjacent span
ℓ_n' = clear span length of shorter adjacent span

For equal adjacent spans, Eq. (13-4) reduces to:

$$M_u = 0.07\ (0.5\ w_\ell \ell_n^2) = 0.035\ w_\ell \ell_n^2$$

Distribute M_u (factored moment) to column above and below floor in proportion to column stiffnesses; generally in proportion to column lengths above and below the floor under consideration.

2.6 LATERAL (WIND) LOAD ANALYSIS

For frames without shear walls, the lateral load effects of wind must be resisted by the "unbraced" frame. For low-to-moderate height buildings, lateral wind analysis of an unbraced frame can be performed by either of two simplified methods: the Portal Method or the Joint Coefficient Method. Both methods can be considered to satisfy the elastic frame analysis requirements of the code. The two methods differ in overall approach. The "Portal Method" considers a vertical slice through the entire building along each row of column lines. The method is well suited to the range of building size and height considered by this manual, particularly those with a regular rectangular floor plan. The "Joint Coefficient Method" considers a horizontal slice through the entire building, one floor at a time. The method can accommodate an irregular floor plan, and provision is made to adjust for a wind loading that is eccentric to the resistance centroid of the floor plan. The JC method considers member stiffnesses, whereas the PM method does not.

The Portal Method is presented in this manual because of its simplicity and its intended application to buildings of regular shape. If a building of irregular floor plan is encountered, the designer is directed to Reference 2.2 for application of the Joint Coefficient Method.

2.6.1 Portal Method

The Portal Method considers a two-dimensional frame bent consisting of a line of columns and their connecting horizontal members (slab-beams) with each frame bent extending the full height of the building. The frame bent is considered to be a series of portal units. Each portal unit consists of two-story height columns with connecting slab-beams. Points of moment contraflexure are assumed at mid-length of both beams and columns. Fig. 2-10 illustrates the portal unit concept applied to the top story of a building frame, with each portal unit separated but acting together.

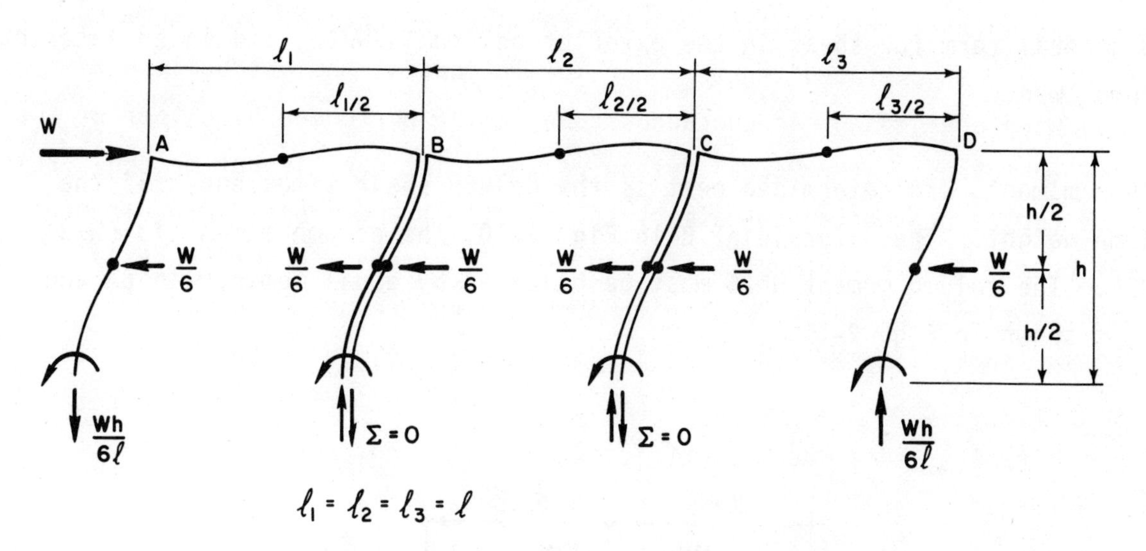

Fig. 2-10 Portal Method

The lateral wind load increment W is divided equally between the three portal units. Consequently, the shear in the interior columns is twice that in the exterior columns. For the case shown with equal spans, axial load occurs only in the exterior columns since the combined tension and compression due to the portal effect results in zero axial load in the interior columns. Under the assumptions of this method, however, a frame configuration with unequal spans will have axial load in those columns between the unequal spans, as well as in the exterior columns. The general terms for axial load in the exterior columns in a frame of n bays with unequal spans is:

$$\frac{Wh}{2n\ell_1} \quad \text{and} \quad \frac{Wh}{2n\ell_n}$$

The axial load in the first interior column is:

$$\frac{Wh}{2n\ell_1} - \frac{Wh}{2n\ell_2}$$

and, in the second interior column:

$$\frac{Wh}{2n\ell_2} - \frac{Wh}{2n\ell_3}$$

The general term for shear in the exterior column is W/2n, and in an interior column, W/n.

Column moments are determined next as the column shear times one-half the column height. Thus, for joint B in Fig. 2-10, the column moment is (W/3) (h/2). The column moment Wh/6 must be balanced by equal moments in BA and BC, as shown in Fig. 2-11.

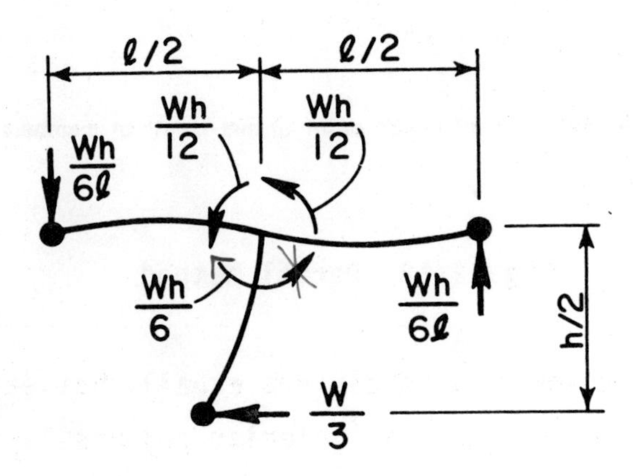

Fig. 2-11 Joint Detail

Note that the balancing moment is divided equally between the horizontal members without considering their relative stiffness. The shear in beams AB and BC is determined next by dividing the beam end moment by one-half the beam length, $(Wh/12)/(\ell/2) = Wh/6\ell$.

The process is continued throughout the frame taking into account the added wind load (story shear) at each floor level.

2.6.1.1 - Example: Lateral Wind Load Analysis by "Portal Method"

For Bldg. #1 (3-story pan joist), determine the moments, shears, and axial loads for an interior frame resulting from wind loads in the NS-direction using the portal method of analysis. Story shears resulting from lateral wind loads are evaluated in Section 2.2.1.2.

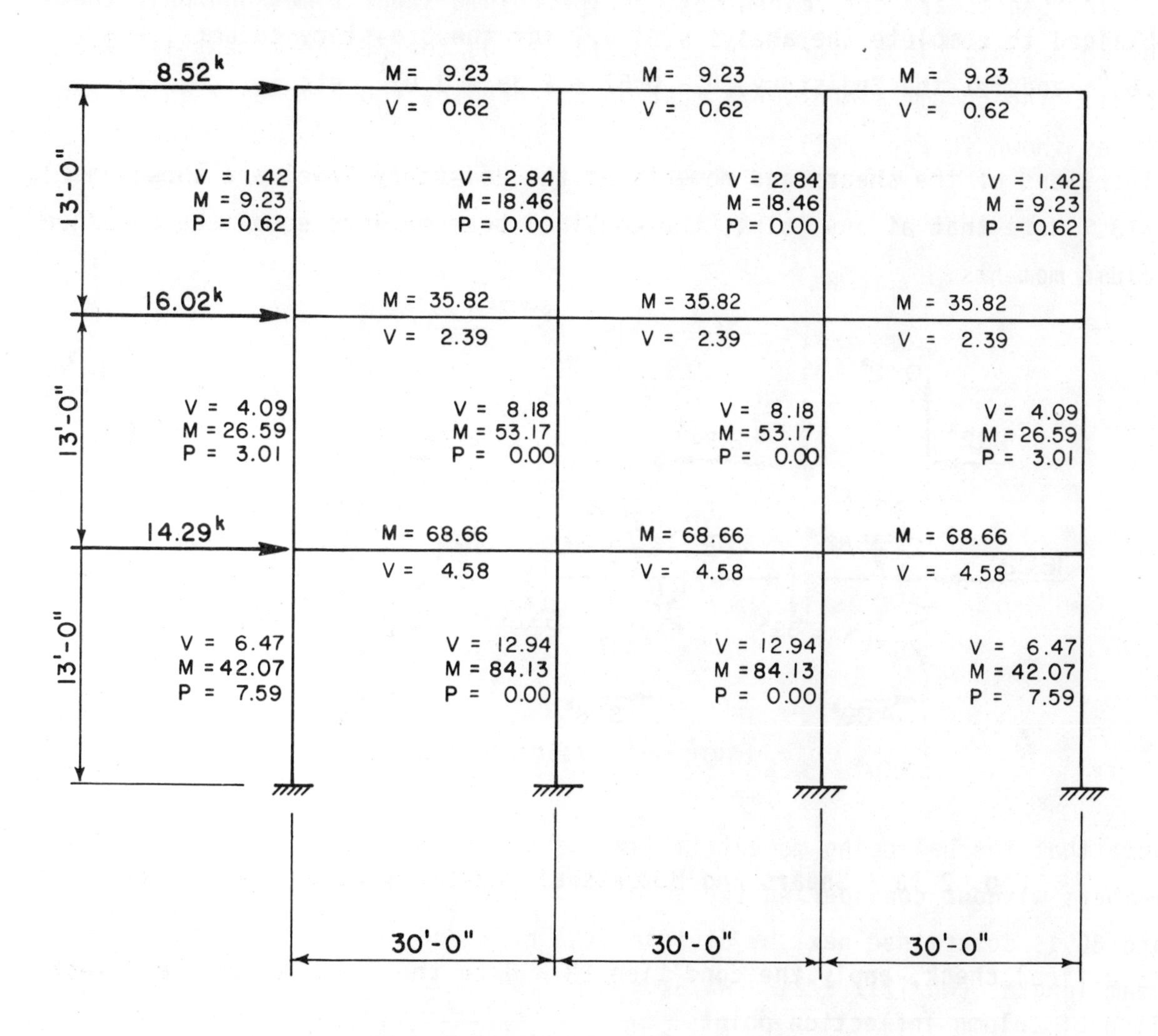

Fig. 2-12 Frame Diagram for Wind Load Analysis (Bldg. #1 - NS-Direction)

Discussion: Moments, shears, and axial loads are developed directly on the frame diagram. The values can be easily checked by noting the order of procedure. The shears V in the columns are recorded first for the entire frame. For the 3rd-story, V = 8.52/3 = 2.84^k; for the 2nd-story, V = (8.52 + 16.02)/3 = 8.18^k; and for the 1st-story, V = (8.52 + 16.02 + 14.29)/3 = ~~19.94~~ 12.94^k. The moments at the top and bottom of each column are next recorded; for the 3rd-story, M = 2.84 x 6.5 = 18.46$^{'k}$. Next, the beam end moments are recorded; M = 18.46/2 = 9.23$^{'k}$. The shears V in the beams are next recorded; V = 9.23/15.0 = 0.62^k. From the beam shears, the column axial loads are obtained to complete the analysis, i.e., for the 3rd-story columns, P = 0.62^k; and for the 2nd-story, P = 0.62 + 2.39 = 3.01^k, etc.

Directions of the shears and moments at the 2nd-story level are shown in Fig. 2-13. Note that at any joint, the combined beam moments equal the combined column moments.

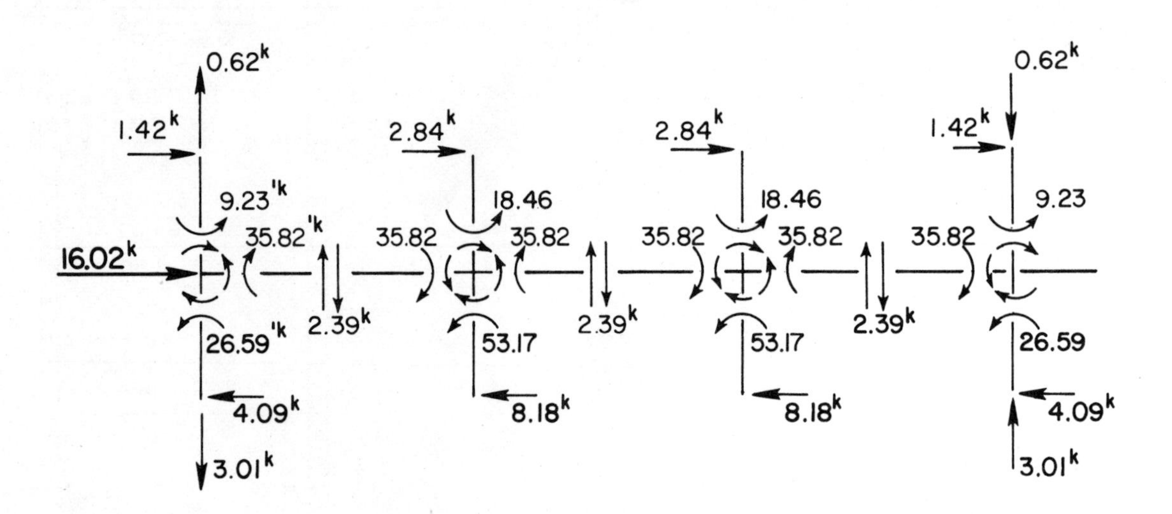

Fig. 2-13 Shears and Moments at 2nd-Story Level - Bldg. #1

As a final check, apply the condition ΣM = 0 to the frame above the lowest line of column inflection points.

In a similar manner, the wind load analysis for an interior frame of Bldg. #2 (5-story flatplate), in both the NS-direction and the EW-direction are shown in Figs. 2-14 and 2-15. Story shears resulting from lateral wind loads are evaluated in Section 2.2.1.1.

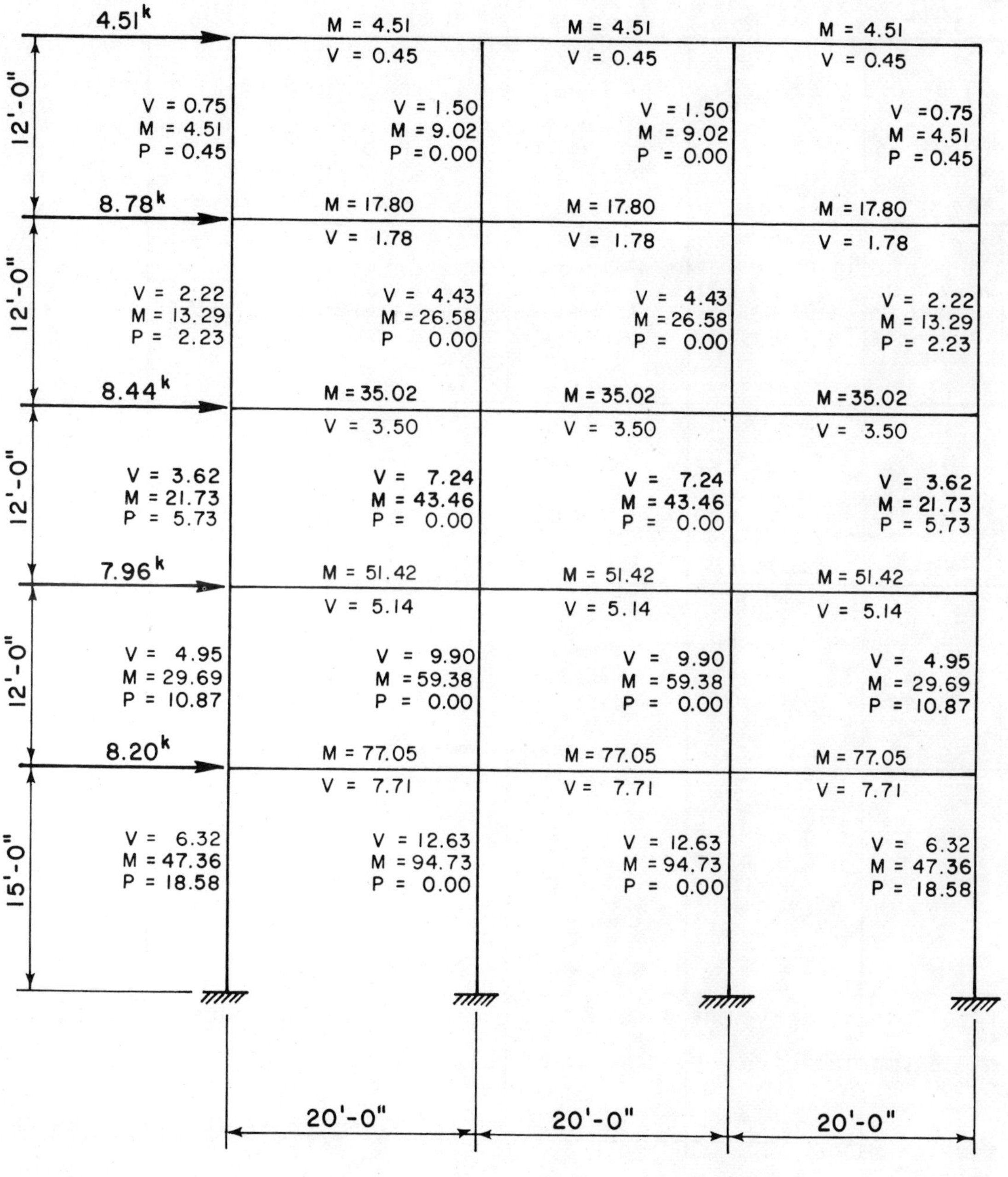

4.51^k

M = 4.51 | M = 4.51 | M = 4.51

V = 0.45 | V = 0.45 | V = 0.45

12'-0"

V = 0.75 | V = 1.50 | V = 1.50 | V = 0.75

M = 4.51 | M = 9.02 | M = 9.02 | M = 4.51

P = 0.45 | P = 0.00 | P = 0.00 | P = 0.45

8.78^k

M = 17.80 | M = 17.80 | M = 17.80

V = 1.78 | V = 1.78 | V = 1.78

12'-0"

V = 2.22 | V = 4.43 | V = 4.43 | V = 2.22

M = 13.29 | M = 26.58 | M = 26.58 | M = 13.29

P = 2.23 | P 0.00 | P = 0.00 | P = 2.23

8.44^k

M = 35.02 | M = 35.02 | M = 35.02

V = 3.50 | V = 3.50 | V = 3.50

12'-0"

V = 3.62 | V = 7.24 | V = 7.24 | V = 3.62

M = 21.73 | M = 43.46 | M = 43.46 | M = 21.73

P = 5.73 | P = 0.00 | P = 0.00 | P = 5.73

7.96^k

M = 51.42 | M = 51.42 | M = 51.42

V = 5.14 | V = 5.14 | V = 5.14

12'-0"

V = 4.95 | V = 9.90 | V = 9.90 | V = 4.95

M = 29.69 | M = 59.38 | M = 59.38 | M = 29.69

P = 10.87 | P = 0.00 | P = 0.00 | P = 10.87

8.20^k

M = 77.05 | M = 77.05 | M = 77.05

V = 7.71 | V = 7.71 | V = 7.71

15'-0"

V = 6.32 | V = 12.63 | V = 12.63 | V = 6.32

M = 47.36 | M = 94.73 | M = 94.73 | M = 47.36

P = 18.58 | P = 0.00 | P = 0.00 | P = 18.58

20'-0" 20'-0" 20'-0"

Fig. 2-14 Frame Diagram for Wind Load Analysis
(Bldg. #2 - NS-direction)

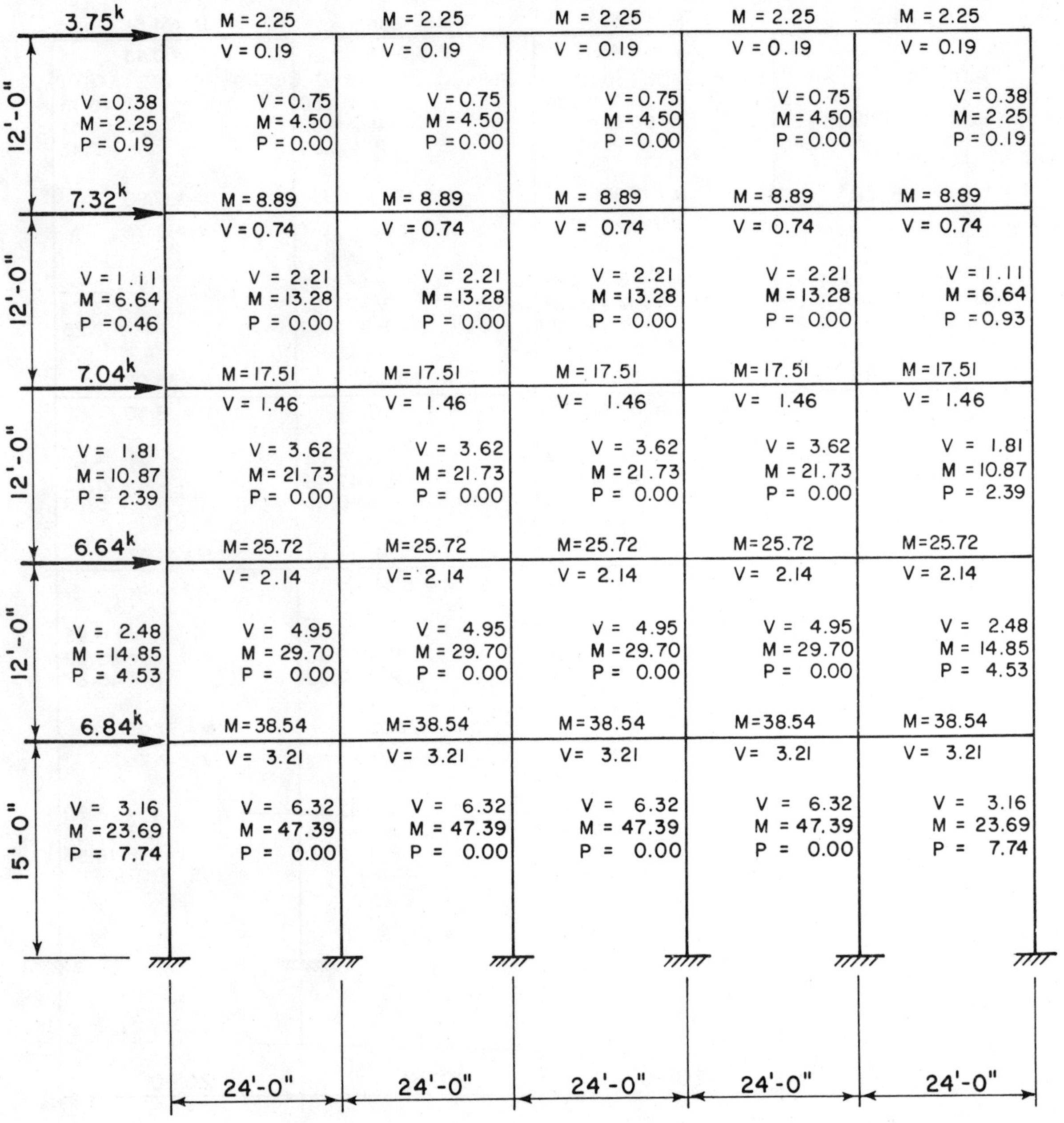

Load	Height	Bay 1	Bay 2	Bay 3	Bay 4	Bay 5
3.75^k		M = 2.25 V = 0.19	M = 2.25 V = 0.19	M = 2.25 V = 0.19	M = 2.25 V = 0.19	M = 2.25 V = 0.19
7.32^k		M = 8.89 V = 0.74	M = 8.89 V = 0.74	M = 8.89 V = 0.74	M = 8.89 V = 0.74	M = 8.89 V = 0.74
7.04^k		M = 17.51 V = 1.46	M = 17.51 V = 1.46	M = 17.51 V = 1.46	M = 17.51 V = 1.46	M = 17.51 V = 1.46
6.64^k		M = 25.72 V = 2.14	M = 25.72 V = 2.14	M = 25.72 V = 2.14	M = 25.72 V = 2.14	M = 25.72 V = 2.14
6.84^k		M = 38.54 V = 3.21	M = 38.54 V = 3.21	M = 38.54 V = 3.21	M = 38.54 V = 3.21	M = 38.54 V = 3.21

Story height	Column 1	Column 2	Column 3	Column 4	Column 5	Column 6
12'-0"	V = 0.38 M = 2.25 P = 0.19	V = 0.75 M = 4.50 P = 0.00	V = 0.75 M = 4.50 P = 0.00	V = 0.75 M = 4.50 P = 0.00	V = 0.75 M = 4.50 P = 0.00	V = 0.38 M = 2.25 P = 0.19
12'-0"	V = 1.11 M = 6.64 P = 0.46	V = 2.21 M = 13.28 P = 0.00	V = 2.21 M = 13.28 P = 0.00	V = 2.21 M = 13.28 P = 0.00	V = 2.21 M = 13.28 P = 0.00	V = 1.11 M = 6.64 P = 0.93
12'-0"	V = 1.81 M = 10.87 P = 2.39	V = 3.62 M = 21.73 P = 0.00	V = 3.62 M = 21.73 P = 0.00	V = 3.62 M = 21.73 P = 0.00	V = 3.62 M = 21.73 P = 0.00	V = 1.81 M = 10.87 P = 2.39
12'-0"	V = 2.48 M = 14.85 P = 4.53	V = 4.95 M = 29.70 P = 0.00	V = 4.95 M = 29.70 P = 0.00	V = 4.95 M = 29.70 P = 0.00	V = 4.95 M = 29.70 P = 0.00	V = 2.48 M = 14.85 P = 4.53
15'-0"	V = 3.16 M = 23.69 P = 7.74	V = 6.32 M = 47.39 P = 0.00	V = 6.32 M = 47.39 P = 0.00	V = 6.32 M = 47.39 P = 0.00	V = 6.32 M = 47.39 P = 0.00	V = 3.16 M = 23.69 P = 7.74

24'-0" 24'-0" 24'-0" 24'-0" 24'-0"

Fig. 2-15 Frame Diagram for Wind Load Analysis
(Bldg. #2 - EW-direction)

Selected References

2.1 "Minimum Design Loads for Buildings and Other Structures," (ANSI A58.1-1982), American National Standards Institute, New York, 1982, 100 pp.

2.2 "Continuity in Concrete Building Frames," Portland Cement Association, Skokie, EB033D, 1959, 56 pp.

3

Simplified Design for Beams and Slabs

Gerald B. Neville*

3.1 - INTRODUCTION

The simplified design approach for proportioning beams and slabs (floor and roof members) is based in part on published articles[3.1,3.2,3.3] suggesting simplified solutions to various conditions of design, and in part on simplified design aid material published by CRSI.[3.4] Additional data for design simplification is added where necessary to provide the designer with a total simplified design approach for beam and slab members. The design conditions that need to be considered for proportioning the beams and slabs are presented in the order generally used in the design process.

The simplified design procedures comply with ACI 318 code requirements for both member strength and member serviceability. The simplified methods will produce slightly more conservative designs within the limitations noted. All coefficients are based on the Strength Design Method, using appropriate load factors and strength reduction factors specified in ACI 318. Where simplified design requires consideration of material strengths, 4000 psi concrete and Grade 60 reinforcement are used. The designer can easily modify the data for other material strengths.

The following design data for f'_c = 4000 psi and f_y = 60,000 psi is easily remembered:

* Manager, Structural Codes, Codes and Standards Department, Portland Cement Association.

modulus of elasticity for concrete	E_c	= 3,600,000 psi	(ACI 8.5.1)
modulus of elasticity for rebars	E_s	= 29,000,000 psi	(ACI 8.5.2)
balanced reinforcement ratio	ρ_b	= 0.0285	(ACI 10.3.2)
minimum reinforcement ratio	ρ_{min}	= 0.0033	(ACI 10.5.1)
maximum reinforcement ratio	ρ_{max}	= $0.75\rho_b$ = 0.0214	(ACI 10.3.3)
temperature reinforcement ratio	ρ_t	= 0.0018	(ACI 7.12.2)

3.2 - DEPTH SELECTION FOR CONTROL OF DEFLECTIONS

Deflection of beams and one-way slabs need not be computed if the overall member thickness meets the minimum specified in ACI Table 9.5(a). Table 9.5(a) may be simplified to four values as shown in Table 3-1; where ℓ_n is length of clear span for cast-in-place beam and slab construction. For design convenience, minimum thickness for the four conditions is plotted in Fig. 3-1.

Table 3-1 - Minimum Thickness for Beams and One-way Slabs

Beams and One-way Slabs	Minimum h
Simple Span Beams or Joists*	$\ell_n/16$
Continuous Beams or Joists	$\ell_n/18.5$
Simple Span Slabs*	$\ell_n/20$
Continuous Slabs	$\ell_n/24$

*Minimum thickness for cantilevers can be considered equal to twice that for a simple span.

Deflections are not likely to cause problems when member overall thickness meets or exceeds these values for uniform loads commonly used in design of buildings. The values are based primarily on experience and are not intended to apply for special cases such as heavy wall or concentrated loadings. Also, they are not intended to apply to members supporting or attached to nonstructural elements likely to be damaged by deflections (ACI 9.5.2.1). For roof beams and slabs, the values are intended for roofs subjected to normal snow or construction live loads only, and with minimal water ponding or drifting (snow) problems.

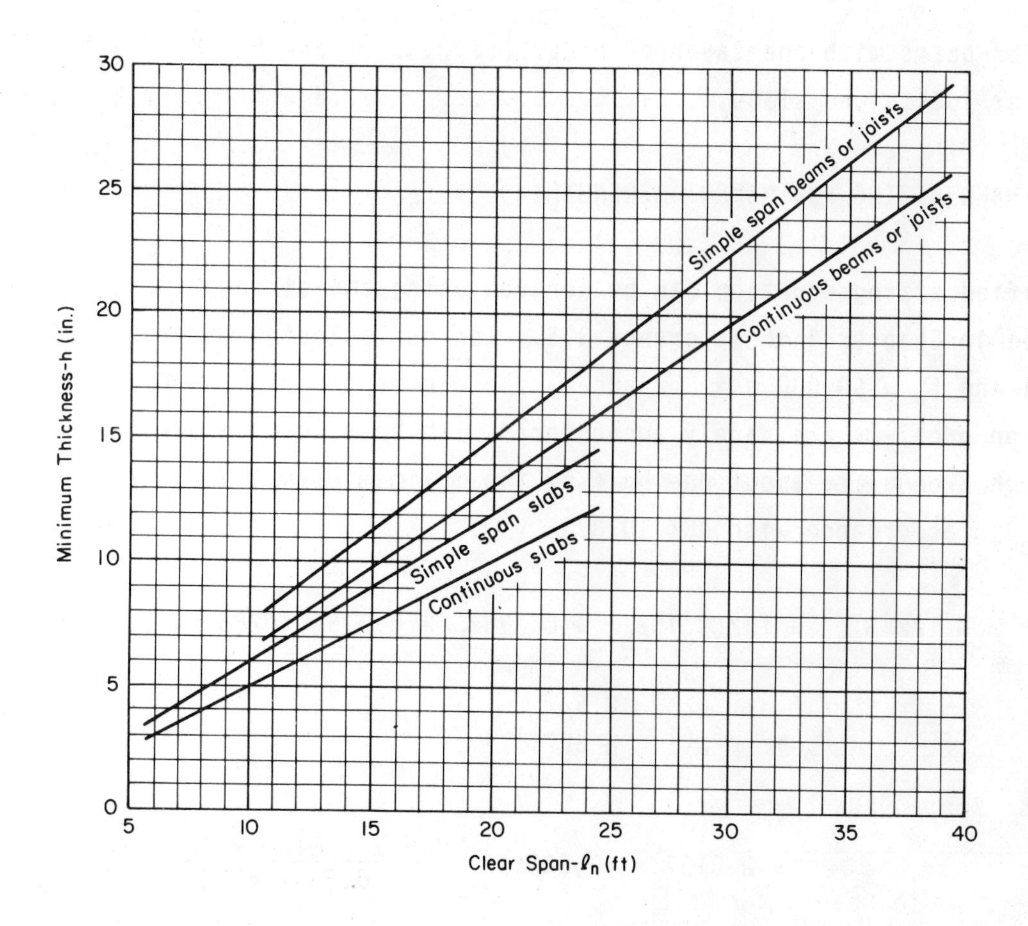

Fig. 3-1 Minimum Thickness for Beams and One-way Slabs

Prudent choice of steel percentage can also minimize deflection problems. Members will usually be of sufficient size so that deflections will be within acceptable limits when the tension reinforcement ratio ρ used in the positive moment regions does not exceed approximately one-half of the maximum permitted

($\rho \simeq 1/2\rho_{max}$). For f'_c = 4000 psi and f_y = 60,000 psi, one-half of ρ_{max} is approximately one percent ($\rho \simeq 0.01$).

Depth selection for control of deflections of two-way slabs is given in Chapter 4.

Using the overall depth h, an effective depth d is calculated in the usual manner. As a guide:

For beams with one layer of bars..............d = h - (≃ 2.5 in.)
For joists and slabs..........................d = h - (≃ 1.25 in.)

3.3 - MEMBER SIZING FOR MOMENT STRENGTH

A simplified sizing equation can be derived using the strength design data developed in Chapter 9 of Reference 3.5. For our selected materials, f'_c = 4000 psi and f_y = 60,000 psi, balanced reinforcement ratio ρ_b= 0.0285. Deflection problems are rarely encountered with beams having a reinforcement percentage ρ equal to about one-half of the maximum permitted. With ρ_{max} = $0.75\rho_b$ in accordance with ACI 10.3.3:

$$\text{Set } \rho = 1/2\ \rho_{max} = 1/2(0.75\rho_b) = 0.375\rho_b = 0.375(0.0285) = 0.0107$$

$$R_n = \rho f_y\left(1 - \frac{0.5\rho f_y}{0.85f'_c}\right)$$

$$= 0.0107 \times 60{,}000\left(1 - \frac{0.5 \times 0.0107 \times 60{,}000}{0.85 \times 4000}\right)$$

$$= 581 \text{ psi}$$

$$bd^2(\text{req'd}) = \frac{M_u}{\varphi R_n} = \frac{M_u \times 12 \times 1000}{0.9 \times 581} = 22.9M_u$$

For simplicity, say $20M_u$

For $f'_c = 4000$ and $f_y = 60{,}000$:

$$bd^2_{(\rho \simeq 1/2\rho_{max})} = 20M_u$$

A similar sizing equation can be derived for other material strengths.

Note: The above sizing equation is in mixed units; M_u in foot-kips and b and d in inches.

With factored moments M_u and effective depth d known, required beam width b is easily determined using the sizing equation $bd^2 = 20M_u$. If frame moments vary widely, a different b may be selected for each beam, or for simplicity, b can be held constant for all similar members of the frame. Since slabs are designed by using a 1-ft strip (b = 12 in.), the sizing equation is used to check the initial depth selected for slabs, which simplifies to $d = 1.3\sqrt{M_u}$.

Also, if the depth determined for control of deflections is shallower than desired, a larger depth may be selected with a corresponding width b determined from the above sizing equation. Actually, any combination of b and d could be determined from the sizing equation with the only restriction being that the final depth selected must be greater than that required for deflection control (Table 3-1).

One final comment on beam sizing...as a check for minimum beam size with maximum reinforcement ($\rho = 0.75\rho_b = 0.0214$), the sizing equation becomes $bd^2_{(min)} = 13M_u$.

3.3.1 - Notes on Member Sizing for Economy

- Use whole inches for overall beam dimensions; slab may be in 1/2-inch increments.

- Use beam widths in multiples of 2 or 3 inches, such as 10, 12, 14, 15, 16, 18, etc.

- Use constant beam size from span to span and vary reinforcement as required.

•Use wide flat beams (same depth as slab system) rather than narrow deep beams.

•Use beam width equal to or greater than column width.

•Use uniform width and depth of beams throughout building.

See also Chapter 9 for design considerations for economical formwork.

3.4 - DESIGN FOR MOMENT REINFORCEMENT

For moment tension reinforcement, a simplified A_s equation can be derived using the strength design data developed in Chapter 9 of Reference 3.5. An approximate linear relationship between R_n and ρ can be described by an equation in the form $M_n/bd^2 = \rho(\text{constant})$, which readily converts to $A_s = M_u/\varphi d(\text{constant})$. This linear equation for A_s is reasonably accurate up to about 2/3 of maximum ρ. For $f'_c = 4000$ and $\rho = 2/3\ \rho_{max}$, the constant for the linear approximation equals 3.93, with M_u in foot-kips and $\varphi = 0.9$ for flexure. For simplicity, use a constant of 4.0.

For $f'_c = 4000$ and $f_y = 60{,}000$:

$$A_s = \frac{M_u}{4d}$$

Note: The above A_s equation is in mixed units; M_u in foot-kips, d in inches and A_s in square inches.

For all values of $\rho < 2/3\ \rho_{max}$, the simplified A_s equation is slightly conservative. The maximum deviation in "A_s" is less than $\pm$ 10% at the minimum and maximum permitted tension steel ratios.[3.6] For members with reinforcement in the range of approximately 1% to 1.5%, the error is less than 3%.

The simplified A_s equation is applicable for rectangular cross sections with tension reinforcement only required for moment strength. Members proportioned with reinforcement in the range of 1% to 1.5% will be well within the code maximum of $0.75\rho_b$ (ACI 10.3.3) for singly reinforced members. For positive moment reinforcement in flanged floor beams, A_s is usually computed as for a rectangular section; rarely will A_s be computed as for a T-shaped compressive zone. Depth "a" as a rectangular section is easily checked by

$$h_f \geq a = A_s f_y / 0.85 f'_c b_e$$

where h_f = thickness of flange (slab thickness)

b_e = effective width of slab as a T-beam flange (ACI 8.10).

3.5 - REINFORCING BAR DETAILS

The minimum and maximum number of reinforcing bars permitted in a given cross section is a function of cover and spacing requirements given in ACI 7.6.1 (minimum spacing for concrete placement), ACI 7.7.1 (minimum cover for protection of reinforcement), and ACI 10.6 (maximum spacing for control of flexural cracking). Tables 3-2 and 3-3 give minimum and maximum number of bars in a single layer for beams of various widths; selection of bars within these limits will provide automatic code conformance for the cover and spacing requirements. The cover and spacing requirements are illustrated in Fig. 3-2. For slabs, maximum spacings to satisfy crack control requirements are stipulated in Table 3-4. Maximum spacings measured center-to-center of parallel bars should not exceed the values in Table 3-4, where h is thickness of slab, flange, or top slab for joist construction. Suggested temperature reinforcement for one-way floor and roof slabs is given in Table 3-5. The required area (per foot width of slab) satisfies ACI 7.12.2. Bar spacing must not exceed 5h or 18 in. The same area of reinforcement also applies for minimum moment reinforcement for one-way slabs (ACI 10.5.3) at a maximum spacing of 3h or 18 in. As noted in Chapter 4, the same A_s applies for minimum moment reinforcement in each direction for two-way floor and roof slabs.....with a maximum spacing of 2h or 18 in.

As an aid to designers, reinforcing bar data are presented in Tables 3-6 and 3-7.

See Chapter 8, Section 8.2, for notes on reinforcement selection and placement for economy.

Table 3-2 - Minimum Number of Bars in Single Layer

Bar Size	INTERIOR EXPOSURE										
	Beam width, b_w (inches)										
	10	12	14	16	18	20	22	24	26	28	30
#5	1	2	2	2	2	2	3	3	3	3	3
#6	1	2	2	2	2	2	3	3	3	3	3
#7	2	2	2	2	2	3	3	3	3	3	4
#8	2	2	2	2	2	3	3	3	3	4	4
#9	2	2	2	2	3	3	3	3	4	4	4
#10	2	2	2	2	3	3	3	3	4	4	4
#11	2	2	2	3	3	3	3	4	4	4	4

Bar Size	EXTERIOR EXPOSURE										
	Beam width, b_w (inches)										
	10	12	14	16	18	20	22	24	26	28	30
#5	2	2	2	2	2	3	3	3	3	3	4
#6	2	3	3	3	4	4	4	5	5	5	6
#7	2	3	3	3	4	4	4	5	5	6	6
#8	2	3	3	4	4	4	5	5	5	6	6
#9	2	3	3	4	4	5	5	5	6	6	7
#10	*	3	3	4	4	5	5	6	6	7	7
#11	*	3	*	4	5	5	5	6	6	7	7

*Maximum permitted is less than minimum number required

Table 3-3 - Maximum Number of Bars in Single Layer

Bar Size	Maximum size coarse aggregate - 3/4 in.										
	Beam width, b_w (inches)										
	10	12	14	16	18	20	22	24	26	28	30
#5	4	5	6	7	8	10	11	12	13	15	16
#6	3	4	6	7	8	9	10	11	12	14	15
#7	3	4	5	6	7	8	9	10	11	12	13
#8	3	4	5	6	7	8	9	10	11	12	13
#9	2	3	4	5	6	7	8	9	9	10	11
#10	2	3	4	5	6	6	7	8	9	10	10
#11	2	3	3	4	5	5	6	7	8	8	9

Bar Size	Maximum size coarse aggregate - 1 in.										
	Beam width, b_w (inches)										
	10	12	14	16	18	20	22	24	26	28	30
#5	3	4	5	6	7	8	9	10	11	12	13
#6	3	4	5	6	7	8	9	9	10	11	12
#7	2	3	4	5	6	7	8	9	10	10	11
#8	2	3	4	5	6	7	7	8	9	10	11
#9	2	3	4	5	6	7	7	8	9	9	10
#10	2	3	4	5	6	6	7	7	8	9	10
#11	2	3	3	4	5	5	6	7	8	8	9

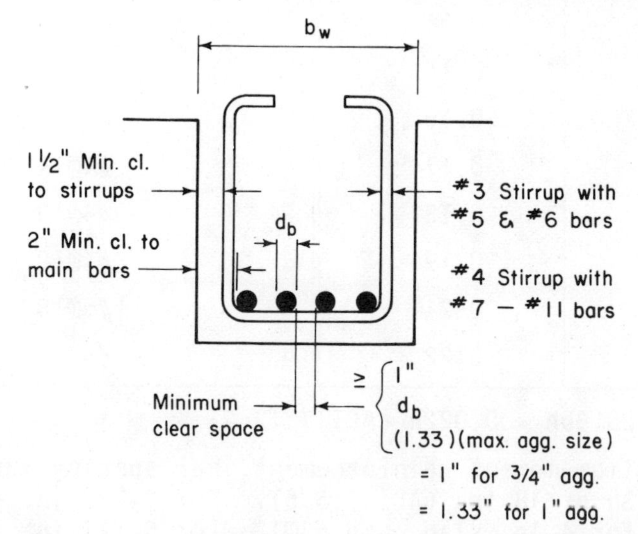

Fig. 3-2 - Cover and Spacing Requirements for Tables 3-2 and 3-3

Table 3-4 - Maximum Bar Spacing for Slabs

Member	Exposure	Bar Sizes	
		#4 - #5	#6 - #11
One-way slabs T-beam flanges and Top slab on joists	Interior	$3h \leq 18"$	$3h \leq 18"$
	Exterior	$3h \leq 8"$	$3h \leq 6"$
Two-way slabs	Interior Exterior	$2h \leq 18"$	

Table 3-5 - Temperature Reinforcement for Slabs

Slab Thickness h	A_s (req'd)* in^2/ft	Suggested Reinforcement**
3-1/2	0.08	#3@16
4	0.09	#3@15
4-1/2	0.10	#3@13
5	0.11	#3@12
5-1/2	0.12	#4@18
6	0.13	#4@18
6-1/2	0.14	#4@17
7	0.15	#4@16
7-1/2	0.16	#4@15
8	0.17	#4@14
8-1/2	0.18	#4@13
9	0.19	#4@12
9-1/2	0.21	#5@18
10	0.22	#5@17

*A_s = 0.0018bh = 0.022h (ACI 7.12.2)

**For minimum moment reinforcement, bar spacing must not exceed 3h or 18 in. (ACI 7.6.5). For 3-1/2 in. slab, use #3@10; for 4 in. slab, use #3@12; for 5-1/2 in. slab, use #3@11 or #4@16.

Table 3-6 - Total Areas of Bars - $A_s(in^2)$

BAR SIZE	BAR DIAMETER	NUMBER OF BARS							
		1	2	3	4	5	6	7	8
# 3	0.375	0.11	0.22	0.33	0.44	0.55	0.66	0.77	0.88
# 4	0.500	0.20	0.40	0.60	0.80	1.00	1.20	1.40	1.60
# 5	0.625	0.31	0.62	0.93	1.24	1.55	1.96	2.17	2.48
# 6	0.750	0.44	0.88	1.32	1.76	2.20	2.64	3.08	3.52
# 7	0.875	0.60	1.20	1.80	2.40	3.00	3.60	4.20	4.80
# 8	1.000	0.79	1.58	2.37	3.16	3.95	4.74	5.53	6.32
# 9	1.128	1.00	2.00	3.00	4.00	5.00	6.00	7.00	8.00
#10	1.270	1.27	2.54	3.81	5.08	6.35	7.62	8.89	10.16
#11	1.410	1.56	3.12	4.68	6.24	7.80	9.36	10.92	12.48

Table 3-7 - Areas of Bars per Foot Width of Slab - A_s (in^2/ft)

BAR SIZE	BAR SPACING - in.												
	6	7	8	9	10	11	12	13	14	15	16	17	18
# 3	0.22	0.19	0.17	0.15	0.13	0.12	0.11	0.10	0.09	0.09	0.08	0.08	0.07
# 4	0.39	0.34	0.29	0.26	0.24	0.22	0.20	0.18	0.17	0.16	0.15	0.14	0.13
# 5	0.61	0.53	0.46	0.41	0.37	0.34	0.31	0.29	0.27	0.25	0.23	0.22	0.21
# 6	0.88	0.76	0.66	0.59	0.53	0.48	0.44	0.41	0.38	0.35	0.33	0.31	0.29
# 7	1.20	1.03	0.90	0.80	0.72	0.65	0.60	0.55	0.51	0.48	0.45	0.42	0.40
# 8	1.57	1.35	1.18	1.05	0.94	0.85	0.78	0.72	0.67	0.62	0.59	0.55	0.52
# 9	2.00	1.71	1.50	1.33	1.20	1.09	1.00	0.92	0.86	0.80	0.75	0.71	0.67
#10	2.53	2.17	1.89	1.69	1.52	1.39	1.27	1.17	1.09	1.02	0.95	0.90	0.85
#11	3.12	2.68	2.34	2.08	1.87	1.70	1.56	1.44	1.34	1.25	1.17	1.10	1.04

3.6 - DESIGN FOR SHEAR REINFORCEMENT

More concerns are expressed about the complexity of shear design provisions than any other provisions of the Code. The confusion perhaps stems from the multitude of strength equations available to compute V_c and V_s. For the majority of design cases, only the simplest of these equations need be applied. Reference 3.3 offers a straightforward simplified method of selecting shear reinforcement.

In accordance with ACI Eq. (11-2) shear strength is made up of two force components: shear strength provided by concrete (φV_c) and shear strength provided by shear reinforcement (φV_s). Thus, $V_u \leq \varphi V_c + \varphi V_s$. Using the simplest of the code equations for shear strength, specific values can be assigned to the two resisting forces for a given set of material parameters and a specific cross section.

The selection and spacing of stirrups can be simplified if the spacing is expressed as a function of the effective depth d instead of numerical values. According to ACI 11.5.4.1 and ACI 11.5.4.3, the practical limits of stirrup spacing vary from s = 1/2d to s = 1/4d, since spacing closer than 1/4d is not economical. With one intermediate spacing at d/3, the calculation and selection of stirrup spacing is greatly simplified. Using only the three standard stirrup spacings (d/2, d/3, and d/4), a specific value of φV_s can be derived for each stirrup size and spacing as follows:

For vertical stirrups:

$$\varphi V_s = \frac{\varphi A_v f_y d}{s} \qquad \text{ACI Eq. (11-17)}$$

substituting d/n for s, where n = 2, 3, or 4:

$$\varphi V_s = \varphi A_v f_y n$$

Thus, for #3 U-stirrups @ s = d/2, f_y = 60,000 and φ = 0.85:

$$\varphi V_s = 0.85(0.22)60 \times 2 = 22.44 \text{ kips, say 22 kips}$$

The following values of φV_s may be used to select shear reinforcement with Grade 60 rebars.

s	φV_s - #3 U-stirrups*	φV_s - #4 U-stirrups*	φV_s - #5 U-stirrups*
d/2	22 kips	40 kips	60 kips
d/3	33 kips	60 kips	90 kips
d/4	44 kips	80 kips	125 kips

*Stirrups with 2 legs, double for 4 legs, etc.

It should be noted that the φV_s values are not dependent on size of member, nor on concrete strength. In a similar manner, design values for φV_c and limiting value for $(\varphi V_c + \varphi V_s)$ can be calculated for $f'_c = 4000$ psi:

$(\varphi V_c + \varphi V_s)$ maximum...........$\varphi 10\sqrt{f'_c} b_w d = 0.55 b_w d$

(φV_c) with stirrups...........$\varphi 2\sqrt{f'_c} b_w d = 0.11 b_w d$

(φV_c) without stirrups..........$\varphi \sqrt{f'_c} b_w d = 0.055 b_w d$

Joists defined by ACI 8.11:
(φV_c) without stirrups.......$\varphi 2.2\sqrt{f'_c} b_w d = 0.12 b_w d$

The selection and spacing of stirrups using the design values for φV_c and φV_s is easily solved either by numerical calculation or graphically.

3.6.1 - Example: Design for Shear Reinforcement

The example shown in Fig. 3-3 illustrates the simple procedure for selecting stirrups using design values for φV_c and φV_s.

(1) Design data: $f'_c = 4000$; $f_y = 60{,}000$, $w_u = 8$ klf

(2) Calculations:

V_u @ column centerline	$w_u \ell/2 = 8 \times 24/2 =$	96.0^k
V_u @ face of support	$96 - 1.17(8) =$	86.6^k
V_u @ "d" distance	$86.6 - 2(8) =$	70.6^k

$(\varphi V_c + \varphi V_s)_{max}$	$0.55 b_w d = 0.55(12)(24) = 158.4^k$
(φV_c) with stirrups	$0.11 b_w d = 0.11(12)(24) = 31.7^k$
(φV_c) without stirrups	$0.055 b_w d = 0.055(12)(24) = 15.8^k$

(3) Beam size is adequate for shear strength, $158.4^k > 70.6^k$.
φV_s(req'd) $= 70.6 - 31.7 = 38.9^k$, #4 @ d/2 is adequate for full length where stirrups are required, $40^k > 38.9^k$. Length over which strirrups are required, $(86.6 - 15.8)/8 = 8.85$ ft
Use <u>10 #4 U-stirrups at 12 in.</u> at each end of beam

Conceivably, the problem can be solved without sketching the shear diagram. For illustration, the problem is solved graphically in Fig. 3-3 for #3 U-stirrups in groups of three spacings. As shown on the shear diagram, φV_s for #3 stirrups at d/2, d/3, and d/4 are scaled vertically from φV_c. The horizontal intersection of the φV_s values (22^k, 33^k, and 44^k) with the shear diagram automatically sets the distances where the #3 stirrups should be spaced at d/2, d/3, and d/4. The exact numerical values for these horizontal distances are calculated as follows, although scaling of the sketch is close enough for practical design.

#3 @ d/4(6")...(86.6 - 64.7)/8 = 2.74'	(32.9")	use 6 @ 6"
@ d/3(8")...(64.7 - 53.7)/8 = 1.38'	(16.6")	use 2 @ 8"
@ d/2(12")..(53.7 - 15.8)/8 = 4.74'	(56.9")	use 5 @ 12"

If #3 stirrups are selected, a more practical solution may be to eliminate the 2 @ 8" spacing and use 9 @ 6" and 5 @ 12".

3.6.2 - Selection of Stirrups for Economy

Selection of stirrup size and spacing for overall cost savings requires consideration of both design time and fabrication and placing costs. An exact solution with an intricate stirrup layout pattern closely following the variation in the shear diagram is not a cost-effective solution. Design time is more cost-effective applied to a quick, more conservative analysis. Small

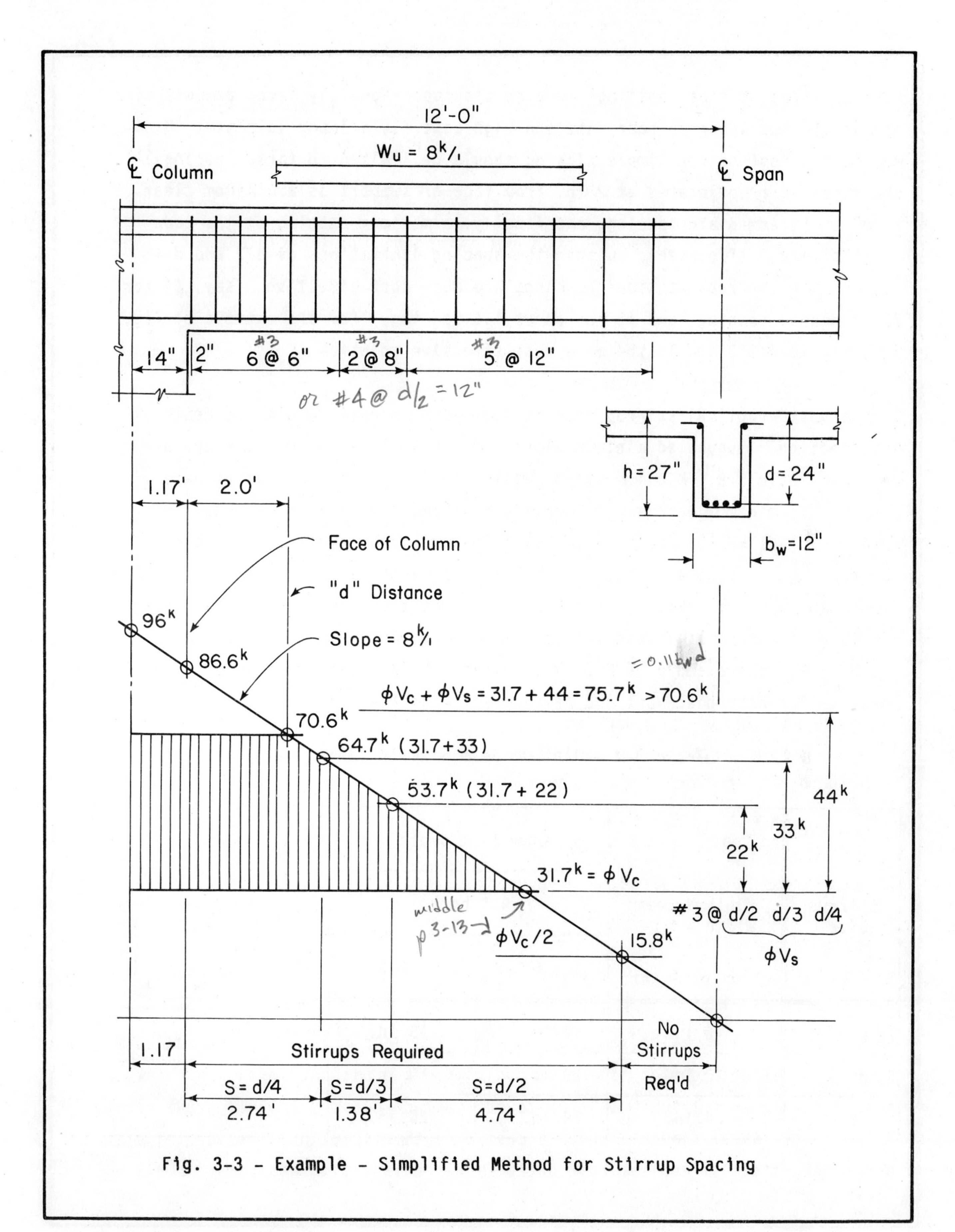

Fig. 3-3 - Example - Simplified Method for Stirrup Spacing

stirrup sizes at close spacings require disproportionately large amounts of high cost shop labor to fabricate and high cost field labor to place. Minimum cost solutions for simple placing should be limited to three spacings: the first stirrup located at 2 in. from face of support as a minimum clearance, an intermediate spacing, and finally, a maximum spacing at the code limit of d/2. If possible, within the spacing limitations of d/2 and d/4, larger size stirrups at wider spacings are more cost-effective...say, #4 for #3 at double spacing, and #5 for #4 at 1.5 spacing. For the example of Fig. 3-3, the 10-#4@12 in. is the more cost-effective solution.

The largest practical stirrup size is limited by overall width and depth of beam section. Suggested minimum width and depth of beams for stirrups are given in Table 3-8. Minimum widths will allow for bend radii at corners of U-stirrups; minimum depths will provide required development of shear reinforcement (ACI 12.13) for f_y = 60 ksi. The required development of the stirrup legs is also a function of stirrup spacing, less than or greater than 6-in. spacing. Note: All stirrups except the first at any group spacing carry substantially less stress than f_y used in calculating minimum depths. For the example of Fig. 3-3, either the #3 or #4 stirrup size is adequate for both placement and development with the 27x12 beam size.

Table 3-8 - Minimum Beam Size for Stirrups

Stirrup Size	Minimum Beam Width - b_w (for Bars in Table 3-3)	Minimum Beam depth - h, 90° hook, s < 6	Minimum Beam depth - h, 90° hook, s ≧ 6
#3	10 in.	17 in.	15 in.
#4	12 in.	20 in.	18 in.
#5	14 in.	24 in.	20 in.

h

b_w

3.7 - DESIGN FOR TORSION

For simplified torsion design of spandrel beams, where the torsional loading is from an integral slab, two options are possible:

(1) Size the spandrel beams so that torsion effects can be neglected (ACI 11.6.1) or,

(2) Provide torsion reinforcement for an assumed maximum torsional moment (ACI 11.6.3).

3.7.1 - Beam Sizing to Neglect Torsion

A simplified sizing equation to neglect torsion effects can be derived based on the limiting factored torsional moment $T_u = \varphi\,(0.5\,\sqrt{f'_c}\,\Sigma x^2 y)$. For a spandrel beam integral with a slab, torsional loading from a slab may be taken as uniformly distributed along the spandrel (ACI 11.6.3.2), with the torsional loading approximated as $t_u = 2T_u/\ell$ per unit length of spandrel beam. With maximum torque T_u at face of support set equal to the limiting torque:

$$t_u = \frac{2\varphi}{\ell}\,(0.5\,\sqrt{f'_c}\,\Sigma x^2 y)$$

For a two-way flat plate or flat slab with spandrel beams, the exterior negative slab moment (ACI 13.6.3.3) is equal to $0.30M_o$, where $M_o = w_u \ell_2 \ell_n^2/8$ (ACI Eq. 13-3). See also Chapter 4, Table 4-3.

Moment transferred to the spandrel per unit length of beam can be approximated as $t_u = 0.30M_o/\ell$. Equating t_u, the required torsional section properties to neglect torsion can be expressed as:

$$\Sigma x^2 y = \frac{0.30M_o}{2\varphi(0.5\,\sqrt{f'_c})}$$

For a spandrel beam section of width b and overall depth h, as shown in Fig. 3-4, the minimum size of spandrel beam to neglect torsion reduces to:

$$b^2h \text{ or } h^2b = \frac{0.30M_o}{2\varphi(0.5\sqrt{f'_c})} - 3h_s^3$$

For our selected concrete strength f'_c = 4000 psi, sizing to neglect torsion simplifies to:

$$\boxed{b^2h \text{ or } h^2b = 67M_o - 3h_s^3}$$

For a one-way slab (or closely spaced beams) with spandrel beam support, the exterior negative slab moment (ACI 8.3.3) is equal to $w_u\ell_n^2/24$, or expressed as a portion of total static span moment, $0.33M_o$. Thus, for a one-way slab with spandrel beams, sizing to neglect torsion reduces to:

$$\boxed{b^2h \text{ or } h^2b = 74M_o - 3h_s^3}$$

The above sizing equations are in mixed units; M_o in foot-kips and b and h in inches.

Architectural or economic considerations may dictate a smaller spandrel size than that required to neglect torsion effects. Additional size of beam to neglect torsion vs. smaller beam size with torsion reinforcement (additional closed stirrups at close spacing combined with longitudinal bars) needs to be evaluated, both from the architectural and economic aspects for a specific floor slab framing system. If a smaller spandrel with torsion reinforcement is a more appropriate choice, Section 3.7.2 provides a simple method for design of torsion reinforcement.

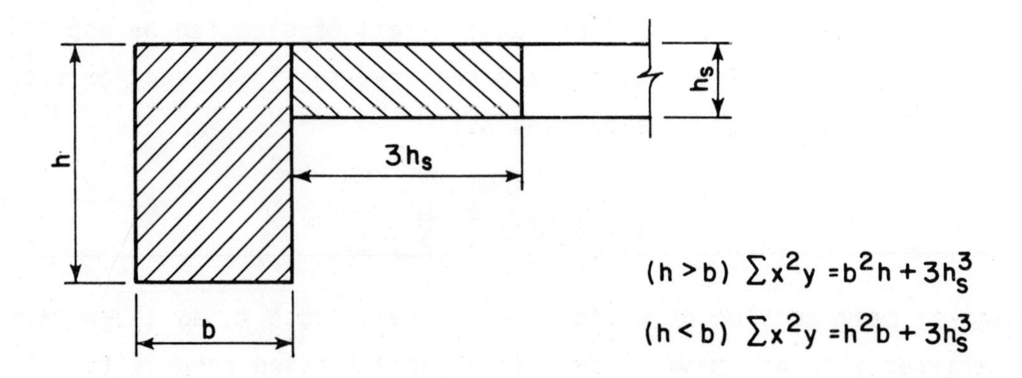

Fig. 3-4 Torsional Section Properties

3.7.1.1 - Example: Beam Sizing to Neglect Torsion

Using the (b^2h) sizing equation, establish a spandrel beam size to neglect torsion effects for Bldg. #2, Alternate (1) - slab and column framing with spandrel beams.

For NS-spandrels: ℓ_2 = 20'-0

ℓ_1 = 24'-0

ℓ_n = 24 - (12+16)/2x12 = 22.83'

w_u = 1.4(136) + 1.7(50) = 275 psf

$$M_o = w_u \ell_2 \ell_n^2/8 = 0.275 \times 20 \times 22.83^2/8 = 358.8^{'k}$$

$$b^2h \text{ or } h^2b_{(req'd)} = 67M_o - 3h_s^3 = 67(358.8) - 3(8.5)^3 = 22{,}197 \text{ in.}^3$$

{ Flat plate
{ w/ spandrel Beam

b	$h_{(req'd)}$	possible selection b x h
12	154	
14	113	
16	87	
18	68.5	
20	55.5	20x56
22	45.9	22x46
24	38.5	24x38
26	32.8	26x34
28	28.3	28x28
30	27.2	30x28
32	26.3	32x26
34	25.6	
36	24.8	

3.7.2 - Simplified Design for Torsion Reinforcement

Torsion reinforcement must consist of a combination of closed stirrups plus longitudinal bars.

For spandrel beams built integrally with a floor slab framing system (where reduction of torsional loading can occur due to redistribution of internal forces), a maximum torsional moment equal to $\varphi\ (4\sqrt{f'_c}\ \Sigma x^2 y/3)$ may be assumed (ACI 11.6.3). Using this condition, simplified design charts for area of closed stirrups (A_t/s) and area of longitudinal bars (A_ℓ) to resist a factored torsional moment of $T_u = \varphi(4\sqrt{f'_c}\ \Sigma x^2 y/3)$ are presented in Figures 3-5 and 3-6. Required A_t/s and A_ℓ are obtained directly as a simple function of spandrel beam width b and overall depth h. Fig. 3-5 is based on ACI Eqs. (11-22) and (11-23) with T_c minimized for maximum A_t/s required. Fig 3-6 is based on ACI Eqs. (11-24) and (11-25) with V_u minimized for maximum A_ℓ required. The torsional section properties of the spandrel include a 5-inch slab (minimum permitted by ACI 9.5.3). Dimensions of the closed stirrup are based on a #3 bar size with 1-1/2-in. cover for concrete exposed to weather... and, of course, our selected materials, $f'_c = 4000$ psi and $f_y = 60{,}000$ psi.

For torsional loading from a uniformly loaded slab to a spandrel beam, the torsional moment variation along the spandrel will be approximately linear from a maximum at the support ($T_u = \varphi(4\ \sqrt{f'_c}\ \Sigma x^2 y/3)$ to zero at midspan. Accordingly, the required area of closed stirrups obtained from Fig. 3-5 may be reduced linearly toward midspan, but not less than the minimum amount required by ACI 11.5.5.5.

One related problem remains for simplified design of spandrel beams for "combined" shear and torsion reinforcement...ACI Eq. (11-5). For $T_u = \varphi(4\sqrt{f'_c}\Sigma x^2 y/3)$, design values for φV_c are listed in Table 3-9 for a range of values for factored shear force V_u expressed as "n" times $\sqrt{f'_c} b_w d$. Use interpolation for intermediate values.

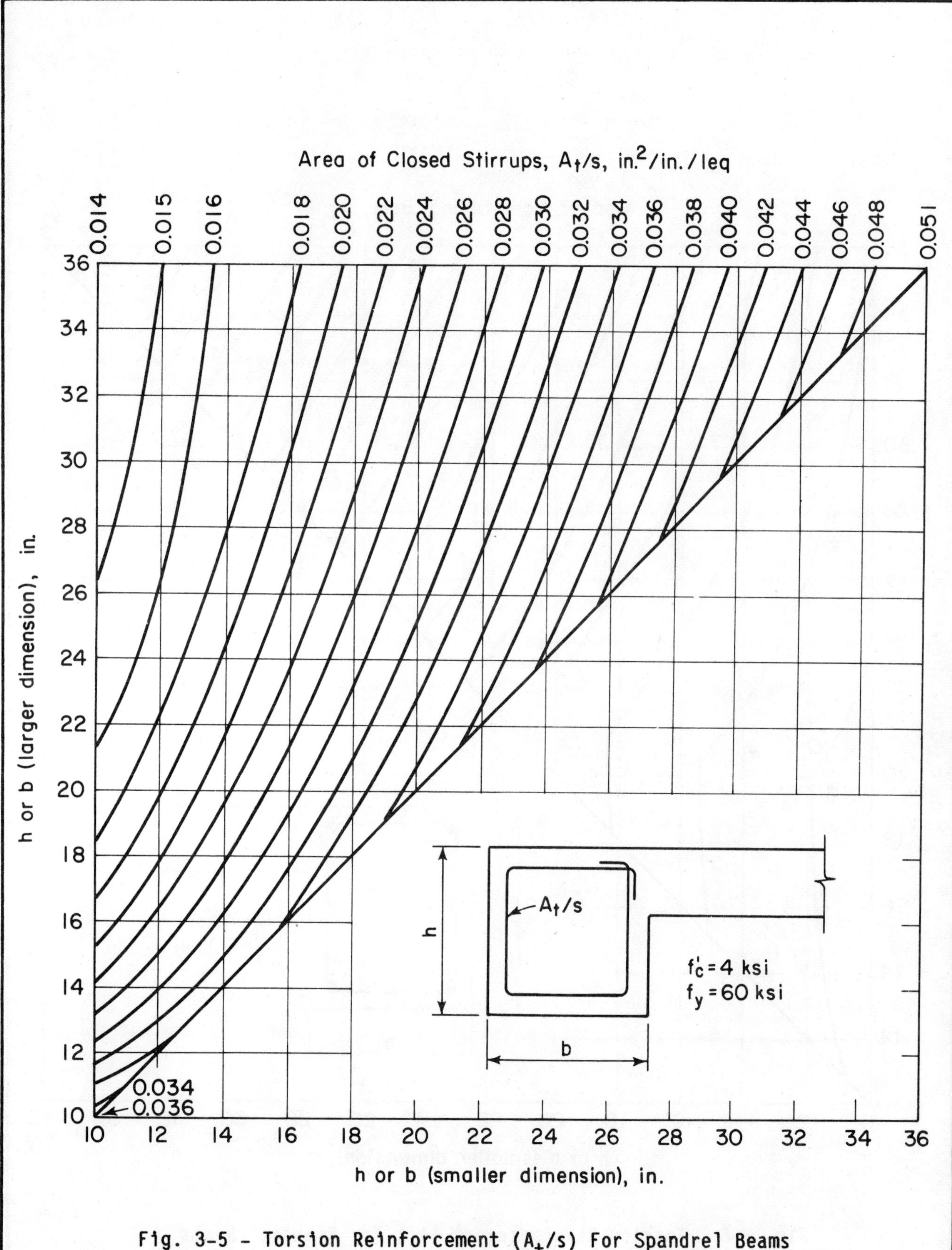

Fig. 3-5 - Torsion Reinforcement (A_t/s) For Spandrel Beams

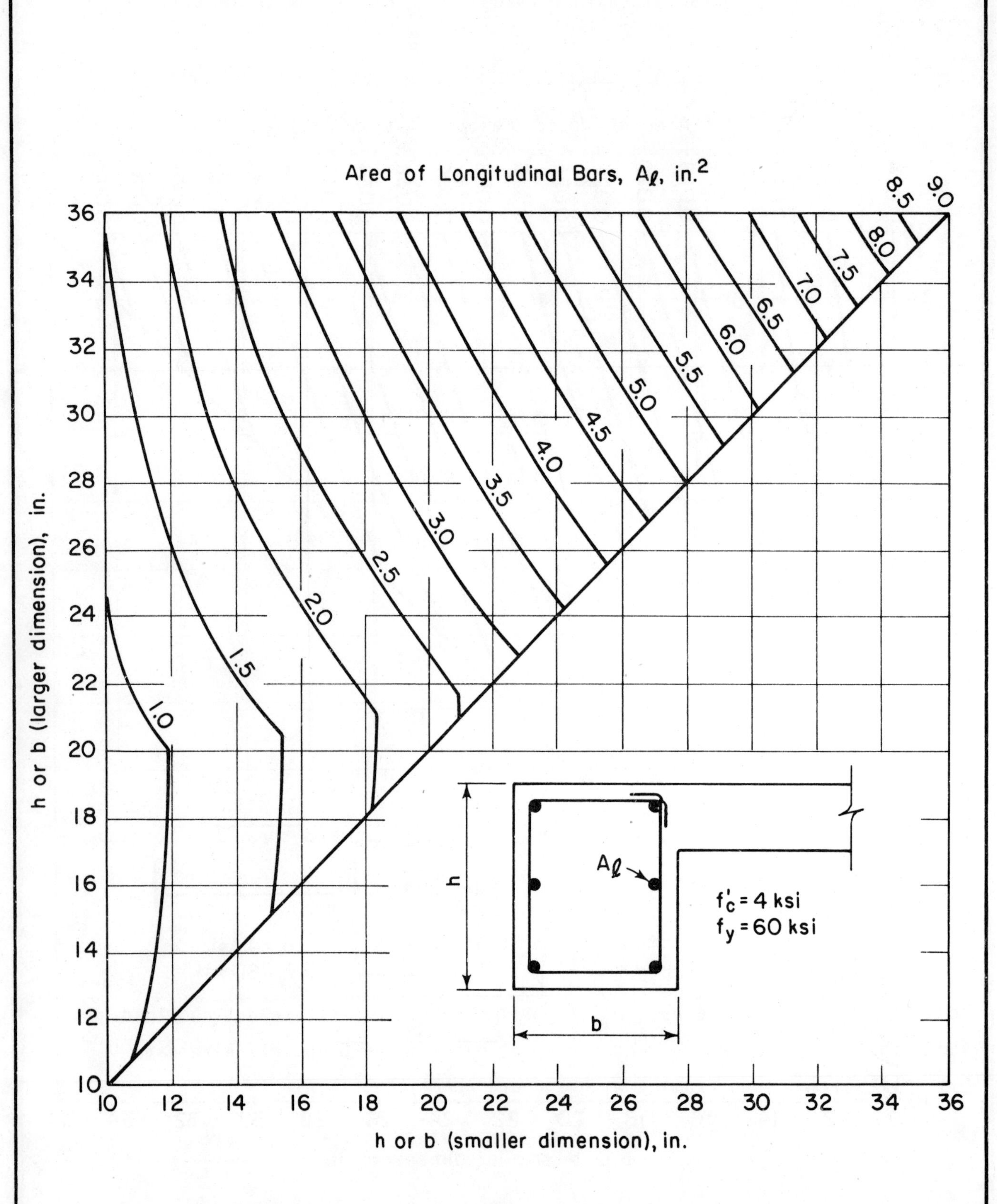

Fig. 3-6 – Torsion Reinforcement (A_ℓ) for Spandrel Beams

Table 3-9 - Shear Strength Provided by Concrete...with Torsion [ACI Eq. (11-5)]

$V_u = n\sqrt{f'_c} b_w d$	(φV_c)
n = 1	$\varphi 0.7\sqrt{f'_c} b_w d = 0.04 b_w d$*
n = 1.5	$\varphi 0.9\sqrt{f'_c} b_w d = 0.05 b_w d$
n = 2.0	$\varphi 1.1\sqrt{f'_c} b_w d = 0.06 b_w d$
n = 2.5	$\varphi 1.3\sqrt{f'_c} b_w d = 0.07 b_w d$
n = 3.0	$\varphi 1.4\sqrt{f'_c} b_w d = 0.075 b_w d$
n = 3.5	$\varphi 1.5\sqrt{f'_c} b_w d = 0.08 b_w d$
n = 4.0	$\varphi 1.6\sqrt{f'_c} b_w d = 0.085 b_w d$
n = 5.0	$\varphi 1.7\sqrt{f'_c} b_w d = 0.09 b_w d$
n = 6.0	$\varphi 1.8\sqrt{f'_c} b_w d = 0.095 b_w d$
$n \geq 7.0$	$\varphi 1.9\sqrt{f'_c} b_w d = 0.10 b_w d$

*f'_c = 4000 psi; φV_c in kips with b_w and d in inches

Example 3.7.2.1 illustrates simplified design for combined shear and torsion reinforcement using Figures 3-5 and 3-6.

3.7.2.1 - Example: Design for Torsion Reinforcement (see p 1-13)

To illustrate simplified design for torsion reinforcement, select combined shear and torsion reinforcement for spandrel beams of Bldg. #2, Alternate (1) - slab and column framing with spandrel beams.

For EW-spandrels:

spandrel size = 12x20

$d = 20-2.5 = 17.5$ in. $= 1.46$ ft

$\ell_n = 24-12/12 = 23.0$

beam wt. $= 1.4\ (12 \times 20 \times 0.150/144) = 0.35^{k/'}$

w_u from slab $= 1.4(136) + 1.7\ (50) = 275$ psf (loads p1-12)

Tributary load to spandrel (1/2 panel width) $= 275(20/2) = 2.75^{k/'}$

beam = 0.35

$w_u = 3.10^{k/'}$

(1) Determine required closed stirrups A_t/s from Fig. 3-5: (p3-21)

For $h = 20$ and $b = 12$; $A_t/s = 0.020$ in.^{2}/in./leg

(2) Calculate required A_v/s:

V_u at face of support $= w_u \ell_n/2 = 3.10 \times 23/2 = 35.7^k$ (24'-1')

V_u at "d" distance $= 35.7 - 1.46\ (3.1) = 31.2^k$

$$n = \frac{V_u}{\sqrt{f'_c} b_w d} = \frac{31.2}{\sqrt{4000} \times 12 \times 17.5} = 2.35$$

From Table 3-9: (p3-23)

$\varphi V_c = \varphi\ 1.24\sqrt{f'_c} b_w d = 0.067 b_w d = 0.067 \times 12 \times 17.5 = 14.1^k$

$\varphi V_s(\text{req'd}) = 31.2 - 14.1 = 17.1^k$

$$A_v/s = \frac{(\varphi V_s)}{\varphi f_y d} = \frac{17.1}{0.85 \times 60 \times 17.5} = 0.019\ \text{in}^2/\text{in.}$$

(3) Determine total stirrup area required for shear and torsion:

The differences in the definitions of A_v and A_t should be carefully noted; A_v is the area of two legs of a closed stirrup, whereas A_t is the area of only one leg of a similar closed stirrup, thus:

$$A_t/s + A_v/2s = 0.020 + 0.019/2 = 0.030 \text{ in.}^2\text{/in./leg}$$

Check minimum stirrup area (ACI 11.5.5.5):

$$A_v/2s + A_t/s = 50b_w/f_y = 50 \times 12/60{,}000 = 0.01 \text{ in.}^2\text{/in./leg} < 0.030 \text{ OK}$$

<u>Try #4 bar</u>, $A_b = 0.20 \text{ in.}^2$

$$s = \frac{0.20}{0.030} = 6.7 \text{ in.}$$

<u>Try #5 bar</u>, $A_b = 0.31 \text{ in.}^2$

$$s = \frac{0.31}{0.030} = 10.3 \text{ in.}$$

(4) Check maximum spacing (ACI 11.6.8.1 and 11.5.4.1):

For #4 bar:

$$x_1 = 12 - 3.50 = 8.5"$$
$$y_1 = 20 - 3.50 = 16.5"$$
$$(x_1 + y_1)/4 = (8.5 + 16.5)/4 = 6.25"$$

For #5 bar:

$$x_1 = 12 - 3.63 = 8.37"$$
$$y_1 = 20 - 3.63 = 16.37"$$
$$(x_1 + y_1)/4 = (8.37 + 16.37)/4 = 6.2"$$
$$d/2 = 17.5/2 = 8.75"$$

<u>Use #4 closed stirrups @ 6 in.</u>

(5) Determine required longitudinal bars A_ℓ from Fig. 3-6:

For h = 20 and b = 12; $A_\ell = 1.05 \text{ in.}^2$

Place longitudinal bars around perimeter of closed stirrups, spaced not more than 12 in., and locate one longitudinal bar in each corner of the closed stirrups (ACI 11.6.8.2). For the 20 in. deep beam, one bar is required at mid-depth on each side face, with 1/3 of total A_ℓ required at top, mid-depth, and bottom of closed stirrups. $A_\ell/3 = 1.05/3 = 0.35$ in.^{2}. Use 2 #4 bars at mid-depth section (one each face). Longitudinal bars required at top and bottom may be combined with the flexural reinforcement.

Details for the shear and torsion reinforcement (at the support) are shown below. The closed stirrups may be reduced toward midspan, but not less than the minimum amount required by ACI 11.5.5.5. The 20 in. deep beam can accommodate the two-piece closed stirrup detail (20-3 = 17 in. > 16 in. lap splice required for #4 bar).

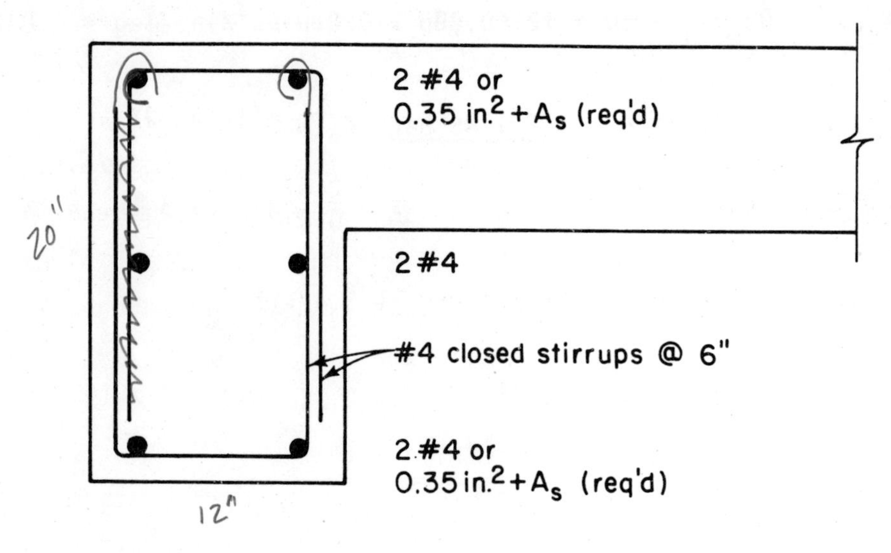

3.7.3 - Closed Stirrup Details

Bar details for closed stirrups are given in ACI 7.11.3. For spandrel beams with torsional loading from an integral slab (low torsion), the two-piece closed stirrup detail illustrated in Fig. 3-7 is suggested. The two U-shaped pieces will facilitate bar placing.

In contrast, the one-piece closed stirrup formed by overlapping 90° end hooks (see Fig. 3-5) has limited bar placing application. The one-piece detail requires either prefabrication of the entire beam cage and placing as a unit

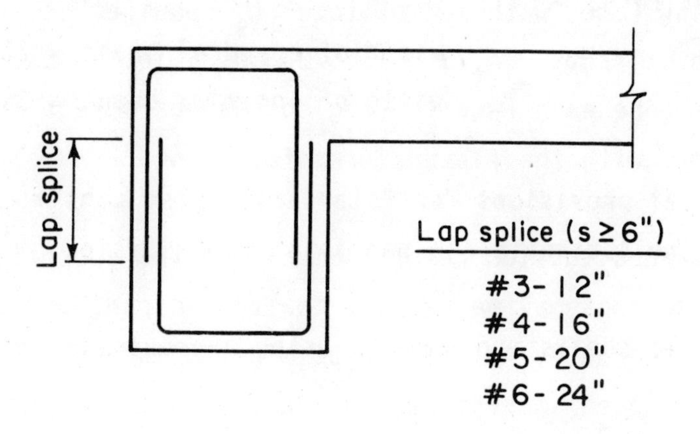

Fig. 3-7 - Two-piece Closed Stirrup Detail

(often impossible because beam bars must pass between column bars) or time consuming assembly in place (perhaps by threading individual beam bars through stirrups and between column bars.) Thus, for practical reasons, the one-piece detail should be used only when bar arrangement and bar placing technique make the detail feasible...or the beam is not deep enough to accommodate the two-piece (lap spliced) detail.

3.8 EXAMPLES: SIMPLIFIED DESIGN FOR BEAMS AND SLABS

The following three examples illustrate use of the simplified design data presented in Chapter 3 for proportioning beams and slabs. Typical floor members for the one-way joist floor system of Bldg. #1 are designed.

3.8.1 - Example: Design Standard Pan Joists for Alternate (1) Floor System

(1) Data: f'_c = 4000 psi (normal weight concrete, carbonate aggregate)
f_y = 60,000 psi (rebars)

Floors: LL = 60 psf
DL = 145 psf (assumed total for joists and beams + partitions + ceiling & misc.)

Fire resistance rating = assume 1 hour

Floor system - Alternate (1): 30-in. standard pan joists
width of spandrel beams = 20 in.
width of interior beams = 36 in.

Note: The special provisions for "standard" joist construction in ACI 8.11 apply for the Alternate (1) pan joist floor system.

(2) Determine factored shears and moments using approximate coefficients of ACI 8.3.3.

Factored shears and moments for the joists of Alternate (1) are determined in Chapter 2, Section 2.3.1.1. Results are summarized as follows:

w_u = 1.4(145) + 1.7(60) = 305 psf x 3 = 915 plf

Note: All shear and negative moment values are at face of supporting beams.

V_u @ spandrel beams	= 12.7^k
V_u @ first interior beams	= 14.5^k
$-M_u$ @ spandrel beams	= $29.0^{'k}$
$+M_u$ @ end spans	= $49.7^{'k}$
$-M_u$ @ first interior beams	= $68.3^{'k}$
$+M_u$ @ interior spans	= $41.9^{'k}$
$-M_u$ @ interior beams	= $61.0^{'k}$

(3) Preliminary size of joist rib and slab thickness

From Table 3-1: depth of joist h = ℓ_n/18.5 = (27.5x12)/18.5 = 17.8 in.
where ℓ_n (end span) = 30 - 1.0 - 1.5 = 27.5 ft (governs)
ℓ_n (interior span) = 30 - 3 = 27.0 ft

From Table 10-1: slab thickness h = 3.2 in. for 1-hour fire resistance rating.

Try 16 in. pan forms* + 3 1/2 in. slab

$h = 19.5 \text{ in.} > 17.8$ OK

*Standard pan forms are 20 and 30 in. widths in depths of 6-8-10-12-14-16-20 in. Tapered end fillers for higher joist shear strength are also available; see Fig. 8-3.

(4) Determine width of joist rib

(a) Code minimum (ACI 8.11.2):

$b_w \geq 16/3.5 = 4.6$ in.

(b) For moment strength:

$$b_w = \frac{20M_u}{d^2} = \frac{20\ (68.3)}{18.25^2} = 4.1 \text{ in.}$$

where $d = 19.5 - 1.25 = 18.25$ in. $= 1.52$ ft

Note: From Tables 10-3 and 10-4, required cover for fire resistance rating of 1 hr = 3/4 in. (both slab and joist).

(c) For shear strength:

V_u @ "d" distance $= 14.5 - 1.52(0.92) = 13.1^k$

φV_c without stirrups $= 1.1(0.11\ b_w d)^* = 0.12\ b_w d$

$b_w = 13.1/0.12 \times 18.25 = 5.98$ in.

Use 6 in. wide joist

*For standard joist ribs, ACI 8.11.8 allows a 10% greater shear strength φV_c. Also, minimum shear reinforcement is not required for joist construction defined by ACI 8.11. See ACI 11.5.5.

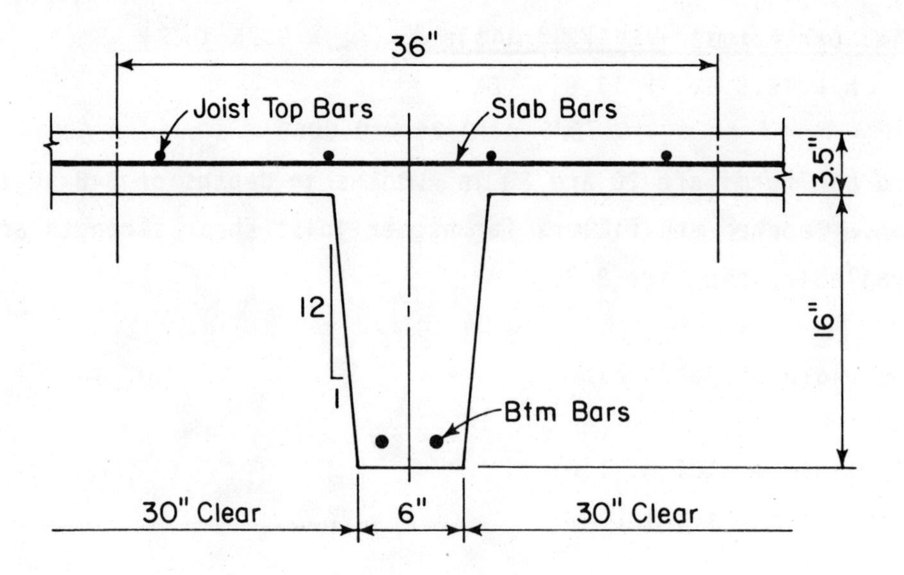

Joist Section

(5) Determine Moment Reinforcement

(a) Top bars at spandrel beams

$$A_s = \frac{M_u}{4d} = \frac{29.0}{4(18.25)} = 0.40 \text{ in.}^2$$

Distribute bars uniformly in top slab:

$$A_s = 0.40/3 = 0.13 \text{ in.}^2\text{/ft}$$

From Table 3-4, maximum bar spacing (interior exposure):

$$s_{max} = 3h = 3(3.5) = 10.5 \text{ in.} < 18 \text{ in.}$$

From Table 3-7: <u>Use #3 @ 10</u> ($A_s = 0.13 \text{ in.}^2\text{/ft}$)

(b) Bottom bars in end spans

$$A_s = 49.7/4(18.25) = 0.68 \text{ in.}^2$$

Check $h_f \geq a = A_s f_y / 0.85 f'_c b_e = 0.68\text{x}60/0.85\text{x}4\text{x}36 = 0.33 \text{ in.} < 3.5 \text{ in.}$ OK

From Table 3-6: Use 1 #5 and 1 #6 (A_s = 0.75 in.^{2})

Check $\rho = A_s/b_w d = 0.75/6 \times 18.25 = 0.0068 > \rho_{min} = 0.0033$ OK

(c) Top bars at first interior beams

$$A_s = 68.3/4(18.25) = 0.94 \text{ in.}^2/3 = 0.31 \text{ in.}^2/\text{ft}$$

From Table 3-7, with s_{max} = 10.5 in.

Use #5 @ 10 (A_s = 0.37 in.^{2}/ft)

(d) Bottom bars in interior spans

$$A_s = 41.9/4(18.25) = 0.57 \text{ in.}^2$$

From Table 3-6: Use 2 #5 (A_s = 0.62 in.^{2})

(e) Top bars at interior beams

$$A_s = 61.0/4(18.25) = 0.84 \text{ in.}^2/3 = 0.28 \text{ in.}^2/\text{ft}$$

From Table 3-7, with s_{max} = 10.5 in.:

Use #5 @ 10 (A_s = 0.37 in.^{2}/ft)

(f) Slab reinforcement normal to ribs (ACI 8.11.6.2)

Slab reinforcement is usually located on the centerline of the slab to resist both positive and negative moments.

Use $M_u = w_u \ell_n^2/12$ (see Fig. 2-5)

$= 0.23(2.5)^2/12 = 0.12'^k$

where $w_u = 1.4(45 + 44) + 1.7(60) = 227 \text{ psf} = 0.23^{k/'}$

ℓ_n = 30 in. (ignore taper)

With bars on slab centerline; $d = 3.5/2 = 1.75$ in.

$$A_s = 0.12/4(1.75) = 0.02 \text{ in.}^2/\text{ft}$$

....but not less than required temperature reinforcement.

From Table 3-5; for 3 1/2 in. slab: <u>Use #3 @ 16</u>

Note: For slab reinforcement normal to ribs, space bars per ACI 7.12.2.2...5h or 18 in.

(7) Bar length details shown below are determined directly from Fig. 8-3.

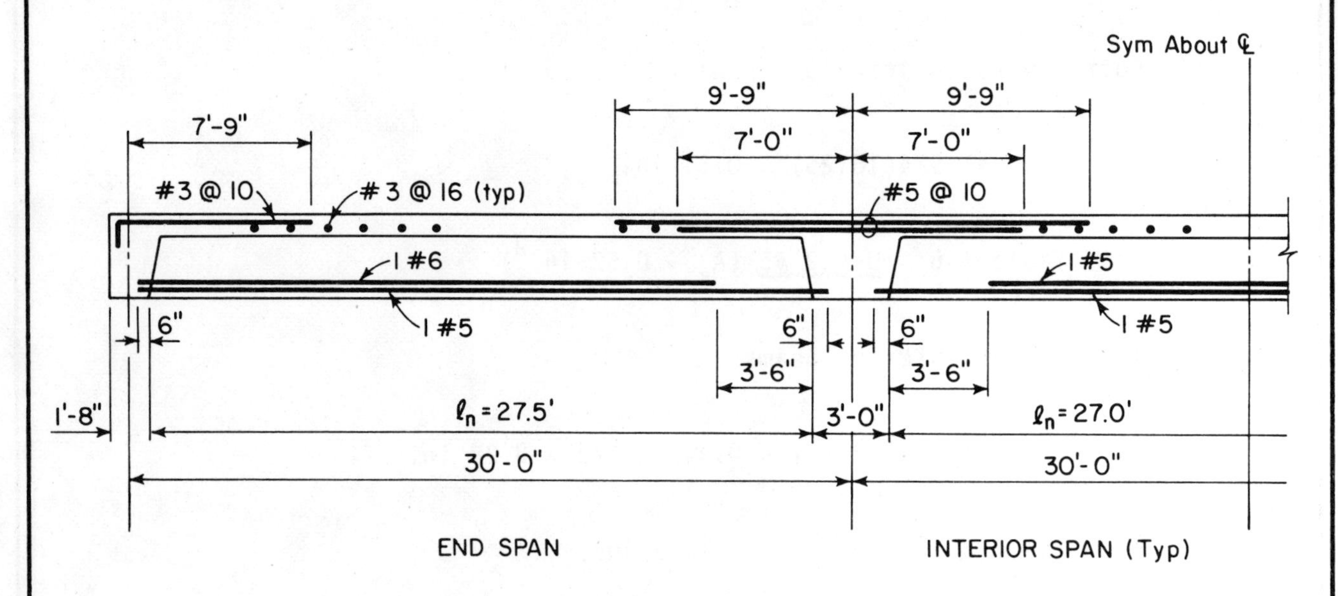

Bar Length Details for 30-in. Standard Joist Floor System
[Bldg. #1 - Alternate (1)]

(8) Distribution Ribs

The ACI Code does not require lateral supports or load distribution ribs in joist construction. Reference 3.7 suggests that distribution ribs 4 in. wide with at least one #4 bar continuous top and bottom be used to equalize deflections and to ensure that the effect of concentrated or unequal loads is resisted by a number of joist ribs. The following spacing for distribution ribs is also recommended:

•None in spans less than 20 ft

•One near the center of spans 20 to 30 ft

•Two near the third points of spans greater than 30 ft.

Provide one distribution rib at center of each span as shown below.

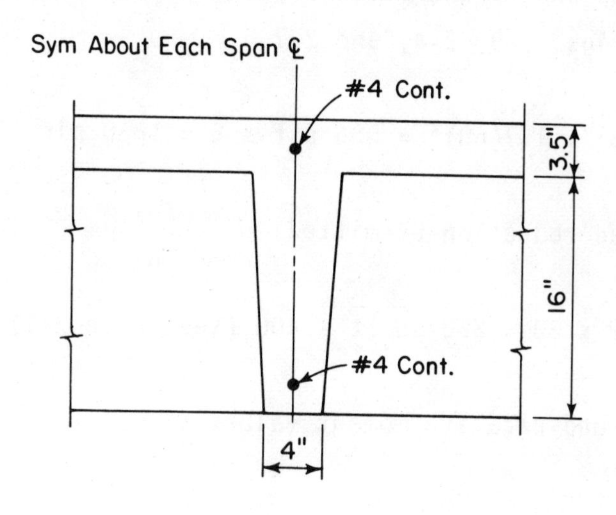

Distribution Rib

3.8.2 - Example: Design Spread Pan Joists for Alternate (2) Floor System

(1) Data: f'_c = 4000 psi (normal weight concrete, carbonate aggregate)
f_y = 60,000 psi (rebars)

Floor: LL = 60 psf
DL = 145 psf (assumed total for beams and slab + partitions + ceiling & misc.)

Fire resistance rating = assume 2 hour

Floor system - Alternate (2): Assume joists on 6-ft centers
width of spandrel beams = 20 in.
width of interior beams = 36 in.

Note: The special provisions for "standard" joist construction in ACI 8.11 do not apply for the spread joist system (clear spacing between joists > 30 in.). Spread joists are designed as beams and one-way slabs (ACI 8.11.4).

(2) Determine factored shears and moments using approximate coefficients of ACI 8.3.3. See Figs. 2-3, 2-4, and 2-7.

$w_u = 1.4(145) + 1.7(60)^* = 305 \text{ psf} \times 6 = 1830 \text{ plf}$

*No live load reduction permitted:

$A_I = 12 \times 30 = 360 \text{ sq ft} < 400$ (see Table 2-1).

Note: All shear and negative moment values are at face of supporting beams.

V_u @ spandrel beams	$= 1.83(27.5)/2 = 25.2^k$
V_u @ first interior beams	$= 1.15(25.2) = 29.0^k$
V_u @ interior beams	$= 1.83(27)/2 = 24.7^k$
$-M_u$ @ spandrel beams	$= 1.83(27.5)^2/24 = 57.7^{'k}$
$+M_u$ @ end spans	$= 1.83(27.5)^2/14 = 98.9^{'k}$
$-M_u$ @ first interior beams	$= 1.83(27.5)^2/10 = 138.4^{'k}$
$+M_u$ @ interior spans	$= 1.83(27)^2/16 = 83.4^{'k}$
$-M_u$ @ interior beams	$= 1.83(27)^2/11 = 121.3^{'k}$

(3) Preliminary size of joist (beam) and slab thickness.

From Table 3-1: depth of joist $h = \ell_n/18.5 = (27.5 \times 12)/18.5 = 17.8$ in.
where ℓ_n (end span) $= 30 - 1.0 - 1.5 = 27.5$ ft (governs)
ℓ_n (interior span) $= 30 - 3 = 27.0$ ft

From Table 10-1: Slab thickness $h = 4.6$ in. for 2-hr rating.

Try 14 in. pan forms + 4 1/2 in. slab

$h = 18.5 \text{ in.} > 17.8$ OK

(4) Determine width of joist rib

(a) For moment strength:

$$b_w = \frac{20M_u}{d^2} = \frac{20(138.4)}{16.5^2} = 10.17 \text{ in.}$$

where d = 18.5 - 2.00 = 16.50 in. = 1.38 ft

Note: For joists designed as beams, minimum thickness of cover = 1 1/2 in. From Tables 10-3 and 10-4, required cover for fire resistance rating of 2 hr = 3/4 in. < 1.5 in.

(b) For shear strength:

V_u @ "d" distance = 29.0 - 1.83(1.38) = 26.5^k

φV_c without stirrups = 0.055 $b_w d$*

*For joists designed as beams, the 10% increase in φV_c is not permitted. Also, minimum shear reinforcement is required when $V_u > \varphi V_c/2$.

$b_w = 26.5/0.055 \times 16.5 = 29.2$ in.

Use 10 in.-wide joist and provide stirrups where required.

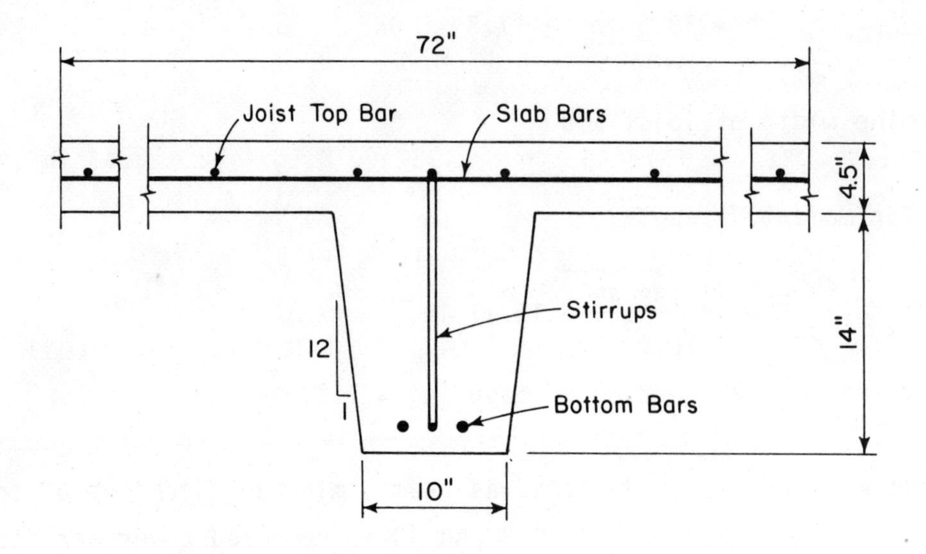

Joist Section

(5) Determine Moment Reinforcement

(a) Top bars at spandrel beams

$$A_s = \frac{M_u}{4d} = \frac{57.7}{4(16.5)} = 0.87 \text{ in.}^2$$

Distribute bars uniformly in top slab according to ACI 10.6.6.

Effective flange width: 30 x 12/10 = 36 in. (governs)

6 x 12 = 72 in.

10 + 2(8 x 4.5) = 82 in.

$$A_s = 0.87/3 = 0.29 \text{ in.}^2/\text{ft}$$

From Table 3-4, maximum bar spacing (interior exposure):

$$s_{max} = 3h = 3(4.5) = 13.5 \text{ in.} < 18 \text{ in.}$$

From Table 3-7: Use #5 @ 13 (A_s = 0.29 in.^{2}/ft)

(b) Bottom bars in end spans

$$A_s = 98.9/4(16.5) = 1.50 \text{ in.}^2$$

Check $h_f \geq a = A_s f_y / 0.85 f'_c b_e = 1.50\text{x}60/0.85\text{x}4\text{x}72 = 0.37 \text{ in.} < 4.5 \text{ in.}$ OK

From Table 3-6: <u>Use 2 #8</u> ($A_s = 1.58 \text{ in.}^2$)

Check $\rho = A_s/b_w d = 1.58/10 \times 16.5 = 0.0096 > \rho_{min} = 0.0033$ OK

The 2 #8 bars are also within the limits of Tables 3-2 and 3-3 for 10 in. wide beam.

(c) Top bars at first interior beams

$$A_s = 138.4/4(16.5) = 2.10 \text{ in.}^2/3 = 0.70 \text{ in.}^2/\text{ft}$$

From Table 3-7, with $s_{max} = 13.5$ in.:

<u>Use #8 @ 13</u> ($A_s = 0.72 \text{ in.}^2/\text{ft}$)

(d) Bottom bars in interior spans

$$A_s = 83.4/4(16.5) = 1.26 \text{ in.}^2$$

From Table 3-6: <u>Use 2 #7</u> ($A_s = 1.20 \text{ in.}^2$)

(e) Top bars at interior beams

$$A_s = 121.3/4(16.5) = 1.84 \text{ in.}^2/3 = 0.61 \text{ in.}^2/\text{ft}$$

From Table 3-7, with $s_{max} = 13.5$ in.

<u>Use #7 @ 12</u> ($A_s = 0.60 \text{ in.}^2/\text{ft}$)

(f) Slab reinforcement normal to joists

Use $M_u = w_u \ell_n^2/12$ (see Fig. 2-5)

$= 0.24(6)^2/12 = 0.72^{'k}$

where $w_u = 1.4(45 + 56) + 1.7(60) = 243$ psf $= 0.24^{k/'}$

Place bars on slab centerline; $d = 4.5/2 = 2.25$ in.

$A_s = 0.72/4(2.25) = 0.08$ in.^{2}/ft

....but not less than required temperature reinforcement.

From Table 3-5; for 4 1/2 in. slab: <u>Use #3 @ 13</u> (A_s = 0.10 in.^{2}/ft)

(6) Bar length details shown below are determined directly from Fig. 8-3.

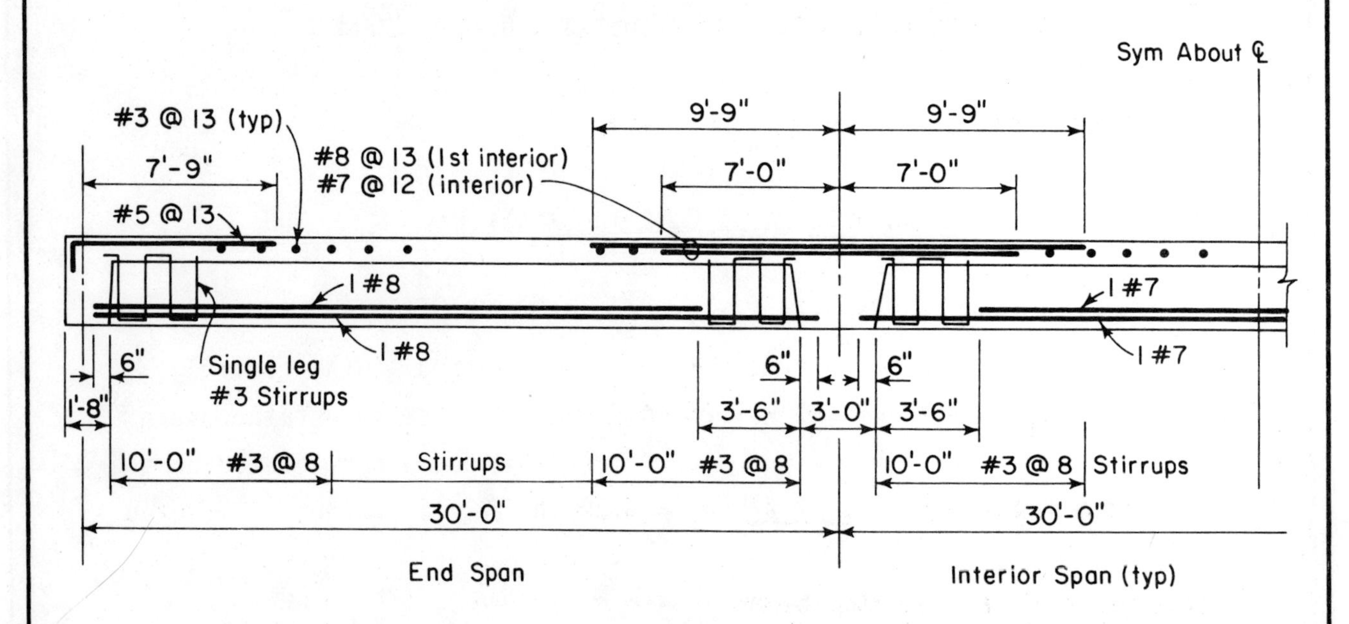

Bar Length Details for 6'-0 Spread Joist Floor System.
[Bldg. #1 - Alternate (2)]

(7) Distribution ribs. Provide one load distribution rib at center of each span.

(8) Design for Shear Reinforcement

(a) End spans

V_u at face of interior beam = 29.0^k

V_u at "d" distance = 29.0 - 1.38(1.83) = 26.5^k
where w_u = 1.83 klf

Use average web width for shear strength calculations.
b_w = 10 + 18.5/12 = 11.54 in.

$(\varphi V_c + \varphi V_s)_{max}$ = 0.55 $b_w d$ = 0.55(11.54)(16.5) = 104.7^k

(φV_c) with stirrups = 0.11 $b_w d$ = 0.11(11.54)(16.5) = 20.9^k

(φV_c) without stirrups = 0.055 $b_w d$ = 0.055(11.54)(16.5) = 10.5^k

Beam size is adequate for shear strength, $104.7^k > 26.5^k$

φV_s(req'd) = 26.5 - 20.9 = 5.6^k

Due to the sloping face of a joist rib and the narrow widths commonly used, shear reinforcement is generally a one-legged stirrup rather than the usual two. The type commonly used is a continuous bar located on the joist centerline and bent into a configuration similar to dental molding.

Use single-leg #3 stirrups, φV_s @ d/2 = $22^k/2 = 11^k > 5.6^k$

Single-leg #3 @ d/2 is adequate for full length where stirrups are required. Length over which stirrups are required, (29.0 - 10.5)/1.83 = 10.1 ft. s = d/2 = 16.5/2 = 8.25 in.

Check $A_{v(min)}$ = 50 $b_w s/f_y$ = 50 x 11.54 x 8.25/60,000 = 0.079 in.^{2} < 0.11
Single-leg #3 OK

Also, joist depth is adequate for development of the #3 stirrups. See Table 3-8: h = 18.5 in. > 15 OK

Use 16 #3 single-leg stirrups @ 8 in. Use same stirrup detail at each end of all joists

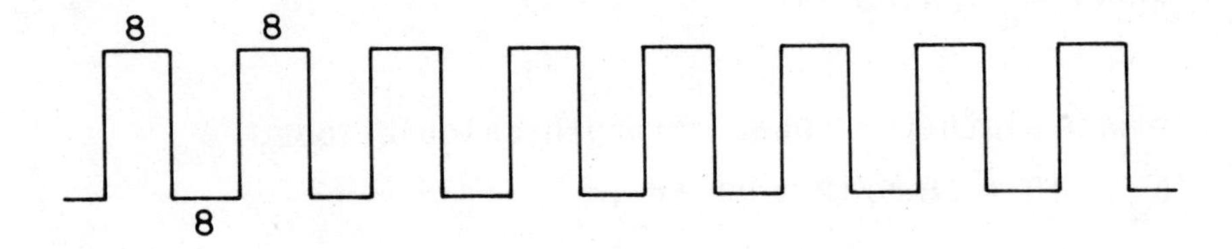

Stirrup Detail

3.8.3 - Design Support Beams for Standard Pan Joist Floor along Typical NS Interior Column Line

(1) Data f'_c = 4000 psi
f_y = 60,000 psi

Floors: LL = 60 psf
DL = 145 psf (assumed total for joists & beams + partitions + ceiling & misc.)

Preliminary member sizes.
Columns int. = 18 x 18
ext. = 16 x 16
width of interior beams = 36 in.

The most economical solution for a pan joist floor is repetition of member size and limiting the depth of supporting beams to the same depth as the joist. In other words, the soffits of the beams and joists will be on a common plane. This reduces form work costs sufficiently to override the savings in materials that may be accomplished by using a deeper member. See Chapter 9 for in-depth discussion on design considerations for economical formwork. Following this suggestion, the beams will often be twice as wide, or more, than they are deep. Overall joist floor depth = 16 in. + 3.5 in. = 19.5 in. Check deflection control for 19.5 in. beam depth; Table 3-1:

$h \geq \ell_n/18.5 = (28.58 \times 12)/18.5 = 18.5 \text{ in.} < 19.5$ OK

where ℓ_n (end span) = 30 - 0.67 - 0.75 = 28.58

ℓ_n (interior span) = 30 - 1.50 = 28.50

(2) Determine factored gravity load shears and moments using approximate coefficients. See Figs. 2-3, 2-4, and 2-7.

Check live load reduction. For interior beams:

$A_I = (30 \times 30)2 = 1800 \text{ sq ft}$

$L_r = 60(0.25 + 15/\sqrt{1800}) = 60(0.604)^* = 36.2 \text{ psf}$

*Note: For members supporting one floor only, maximum reduction = 0.500. See Table 2-1.

$w_u = 1.4(145) + 1.7(36.2) = 265 \text{ psf} \times 30 = 7.95 \text{ klf}$

Note: All shear and negative moment values are at face of supporting columns.

V_u @ exterior columns = $7.95(28.58)/2 = 113.6^k$

V_u @ first interior column = $1.15(113.6) = 130.6^k$

V_u @ interior columns = $7.95(28.5)/2 = 113.3^k$

$-M_u$ @ exterior columns = $7.95(28.58)^2/16 = 405.9^{'k}$

$+M_u$ @ end spans = $7.95(28.58)^2/14 = 463.8^{'k}$

$-M_u$ @ first interior columns = $7.95(28.58)^2/10 = 649.4^{'k}$

$+M_u$ @ interior span = $7.95(28.50)^2/16 = 403.6^{'k}$

(3) Determine wind load moments. Design of the column line beams also includes consideration of lateral load effects of wind. Wind load analysis for Bldg. #1 is summarized in Fig. 2-12.

Note: The 0.75 factor permitted for the load combination including wind effects [ACI Eq. (9-2)] is, in most cases, sufficient to accommodate the wind forces and moments, without an increase in beam size or reinforcement required....load combination for gravity load only will usually govern for proportioning the beams.

(4) Check beam size for moment strength

Preliminary beam size = 19.5 in. x 36 in.

For negative moment section:

$$b = \frac{20M_u}{d^2} = \frac{20(649.4)}{17^2} = 44.9 \text{ in.} > 36 \text{ in.}$$

where d = 19.5 - 2.5 = 17.0 in. = 1.42 ft

For positive moment section:

$$b = 20(463.8)/17^2 = 32.1 \text{ in.} < 36 \text{ in.}$$

Check minimum size permitted with $\rho = 0.75\ \rho_b = 0.0214$:

$$b = 13(649.4)/17^2 = 29.2 \text{ in.} < 36 \text{ in.} \quad \text{OK}$$

Use 36 in. wide beam and provide slightly higher percentage reinforcement ($\rho > 1/2\ \rho_{max}$) at interior columns.

(5) Determine Moment Reinforcement

(a) Top bars at exterior columns.

Check governing load combination....

gravity load only:

$M_u = 405.9^{'k}$ (governs) ACI Eq. (9-1)

gravity + wind loads:

$M_u = 0.75(405.9) + 0.75(1.7 \times 68.7) = 392^{'k}$ ACI Eq. (9-2)

also check for possible moment reversal due to wind moments....

$M_u = 0.9(222.1) \pm 1.3(68.7) = 110.6^{'k}$ ACI Eq. (9-3)

(No reversal)

where $w_d = 145(30) = 4.35$ klf

$M_d = 4.35(28.58)^2/16 = 222.1^{'k}$

$$A_s = \frac{M_u}{4d} = \frac{405.9}{4(17)} = 5.97 \text{ in.}^2$$

From Table 3-6: Use 6 #9 bars ($A_s = 6.00$ in.^{2})

Check $\rho = A_s/bd = 6.00/36 \times 17 = 0.0098 > \rho_{min} = 0.0033$ OK

Projecting the data in Table 3-2 for minimum number of bars (interior exposure) in a 36 in. wide beam, the 6 #9 bars appear adequate for flexural cracking.

(b) Bottom bars in end spans

$A_s = 463.8/4(17) = 6.82$ in.^{2}

Use 7 #9 bars ($A_s = 7.00$ in.^{2})

(c) Top bars at interior columns

Check governing load combination....

gravity load only:

$M_u = 649.4^{'k}$ (governs) Eq. (9-1)

gravity + wind loads:

$M_u = 0.75(649.4) + 0.75(1.7 \times 68.7) = 574.6^{'k}$ Eq. (9-2)

$A_s = 649.4/4(17) = 9.55 \text{ in.}^2$

<u>Use 10 #9 bars</u> ($A_s = 10.0 \text{ in.}^2$)

(d) Bottom bars in interior span

$A_s = 403.6/4(17) = 5.93 \text{ in.}^2$

<u>Use 6 #9 bars</u> ($A_s = 6.00 \text{ in.}^2$)

(6) Bar length details shown below are determined directly from Fig. 8-1. Even though the wind load combination did not govern for required beam reinforcement, provide 2 #5 bars within the center portion of all spans to account for any variations in required bar lengths due to wind effects.

Since the column-line beams are part of the primary wind-force resisting system, ACI 12.11.2 requires at least one-fourth the positive moment reinforcement to be extended into the supporting columns and be anchored to develop full f_y at face of support. For the end spans; $A_s/4$ = 7/4 = 1.75 bars. Extend the 3 #9 center bars anchorage distance into the supports:

- At the exterior columns, provide a 90° standard end-hook. From Table 8-2; for #9 bar, $\ell_{dh} = 15(1.75/3) = 8.75 \text{ in.} < 14 \text{ in.}$ OK

•At the interior columns, extend bar a straight ℓ_d length. From Table 8-1; for the #9 bar, ℓ_d = 38(1.75/3) = 22.2 in. Say 23 - 9 = 1 ft-2 in.

For the interior span; $A_s/4$ = 6/4 = 1.5 bars. Extend the 2 #9 center bars, ℓ_d = 38(1.5/2) = 28.5 in. Say 29 - 9 = 1 ft-8 in.

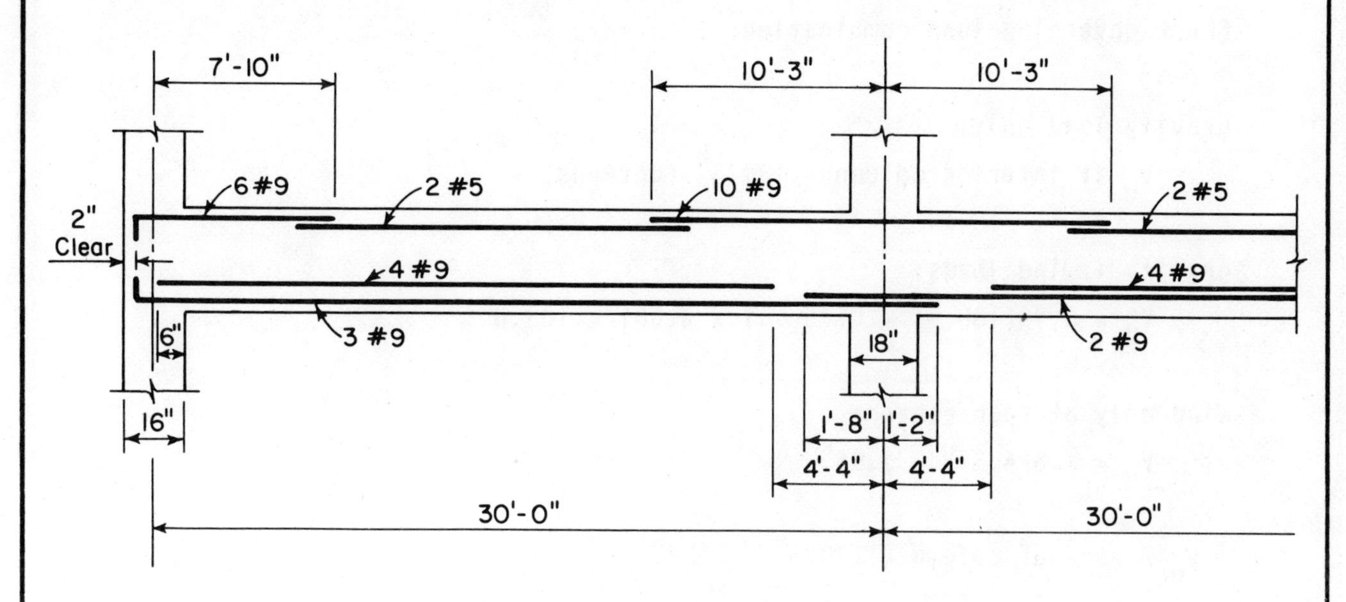

Bar Length Details for Support Beams along NS Interior Column Line

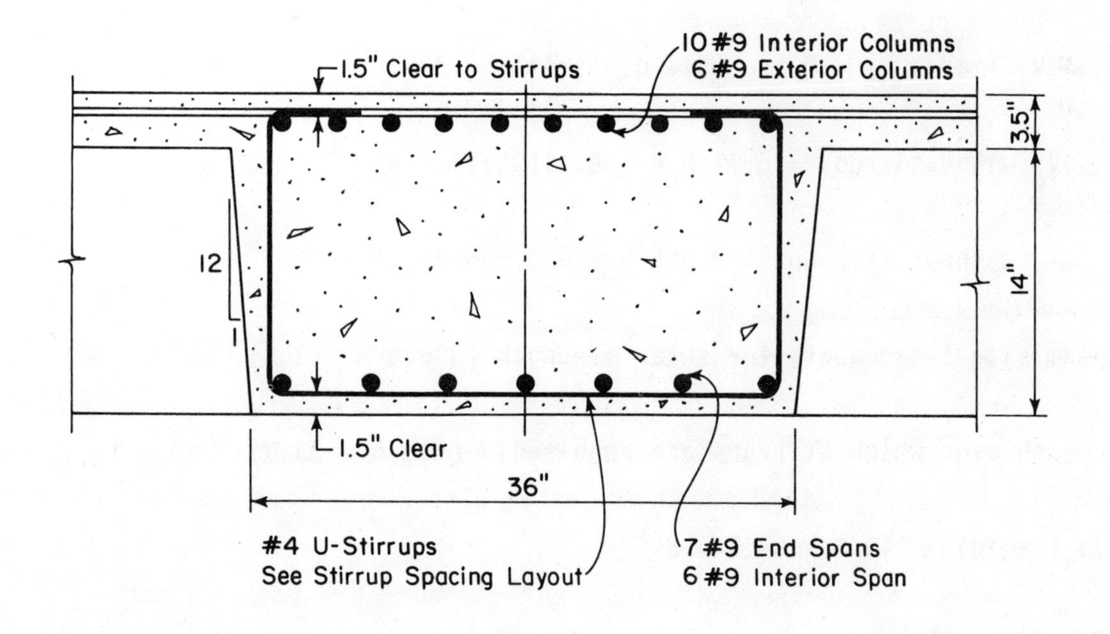

Beam Section

(7) Design for Shear Reinforcement

Design shear reinforcement for end span at interior column and use same stirrup requirements for all three spans.

Check governing load combination....

gravity load only:

V_u at interior column = 130.6^k (governs)

gravity + wind loads:

$V_u = 0.75(130.6) + 0.75(1.7 \times 4.58) = 103.8^k$

wind only at span center:

$V_u = 1.3(4.58) = 5.95^k$

V_u @ face of column = 130.6^k

V_u at "d" distance = $130.6 - 1.42(7.95) = 119.3^k$
where w_u = 7.95 klf

$(\varphi V_c + \varphi V_s)_{max} = 0.55\ b_w d = 0.55(36)17 = 336.6^k$

φV_c with stirrups = $0.11\ b_w d = 0.11(36)17 = 67.3^k$

φV_c without stirrups = $0.055\ b_w d = 0.055(36)17 = 33.7^k$

Beam size is adequate for shear strength (336.6 > 119.3).

Length over which stirrups are required: (130.6 - 33.7)/7.95 = 12.19 ft

φV_s(req'd) = $119.3 - 67.3 = 52^k$

Try #4 U-stirrups

#4 @ d/2, $\varphi V_c + \varphi V_s = 67.3 + 40 = 107.3^k$

#4 @ d/3, $\varphi V_c + \varphi V_s = 67.3 + 60 = 127.3^k$

Length over which #4 @ d/3 required: (130.6 - 107.3)/7.95 = 2.93 ft

spaces @ d/3 (say 6 in.) = 2.93/0.5 = 5.9 <u>use 6 spaces @ 6 in.</u>

Length over which #4 @ d/2 required: 12.19 - 2.93 = 9.26 ft

spaces @ d/2 (say 9 in.) = 9.26/0.75 = 12.3 <u>use 13 spaces @ 9 in.</u>

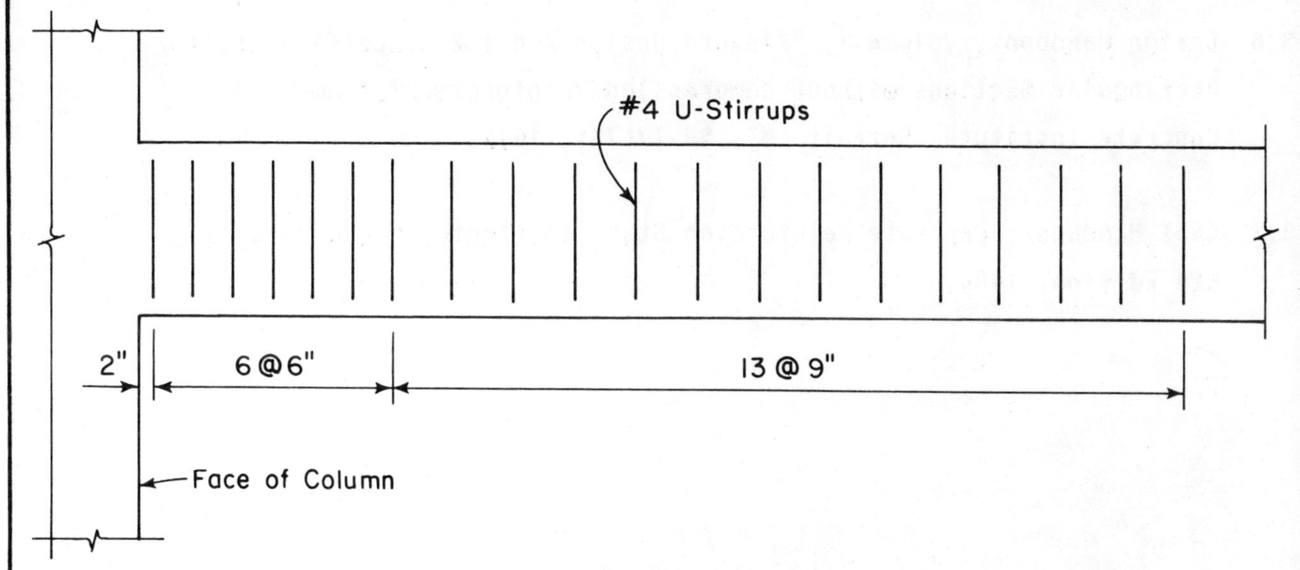

Stirrup Spacing Layout - Use same layout at each end of all beams.

Selected References

3.1 Fling, Russell S., "Using ACI 318 the Easy Way," Concrete International, V. 1, No. 1, January 1979.

3.2 Pickett, Chris, "Notes on ACI 318-77, Appendix A - Notes on Simplified Design," 3rd Edition, Portland Cement Association, Skokie, IL.

3.3 Rogers, Paul, "Simplified Method of Stirrup Spacing," Concrete International, V. 1, No. 1, January 1979.

3.4 "PSI - Product Services and Information," Concrete Reinforcing Steel Institute, Schaumburg, IL.

(a) Bulletin 7901A "Selection of Stirrups in Flexural Members for Economy"

(b) Bulletin 7701A "Reinforcing Bars Required - Minimum vs. Maximum"

(c) Bulletin 7702A "Serviceability Requirements with Grade 60 Bars"

3.5 "Notes on ACI 318-83, Chapter 9, Design for Flexure," 4th Edition, EB070.04D, Portland Cement Association, Skokie, IL.

3.6 Design Handbook, Volume 1, "Flexure Design Aid 1.2 - Coefficients for Rectangular Sections without Compression Reinforcement," American Concrete Institute, Detroit, MI, SP-17(73), 1973.

3.7 CRSI Handbook, Concrete Reinforcing Steel Institute, Schaumburg, IL, 6th Edition, 1984.

4

Simplified Design for Two-Way Slabs

S. K. Ghosh*

4.1 INTRODUCTION

Figure 4-1 shows the various types of two-way reinforced concrete slab systems in use at the present time.

A solid slab supported on beams on all four sides [Fig. 4-1(a)] was the original slab system in reinforced concrete. With this system, if the ratio of the long to the short side of a slab panel is two or more, load transfer is predominantly by bending in the short direction and the panel essentially acts as a one-way slab. As the ratio of the sides of a slab panel approaches unity (square panel), significant load is transferred by bending in both orthogonal directions, and the panel should be treated as a two-way rather than a one-way slab.

As time progressed and technology evolved, the column-line beams gradually began to disappear. The resulting slab system consisting of solid slabs supported directly on columns is called the flat plate [Fig. 4-1(b)]. The two-way flat plate is very efficient and economical and is currently the most widely used slab system for multistory residential and institutional construction, such as motels, hotels, dormitories, apartment buildings, and hospitals. In comparison to other concrete floor/roof systems, flat plates

---

*Associate Professor of Civil Engineering, University of Illinois at Chicago.

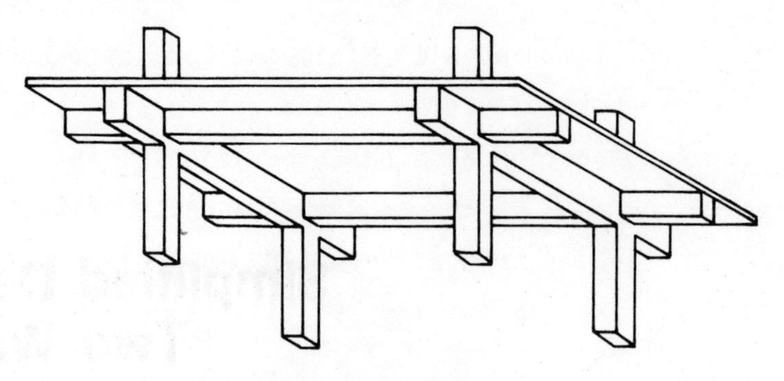

(a) Two-way Slab

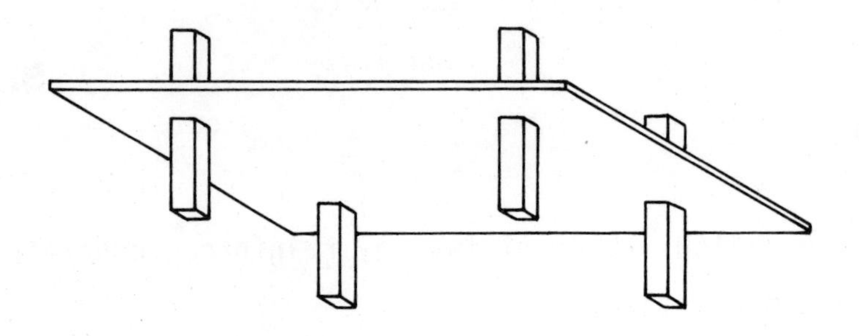

(b) Flat Plate

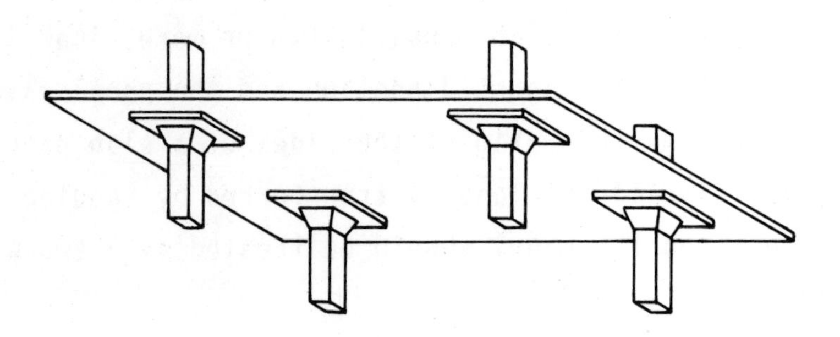

(c) Flat Slab

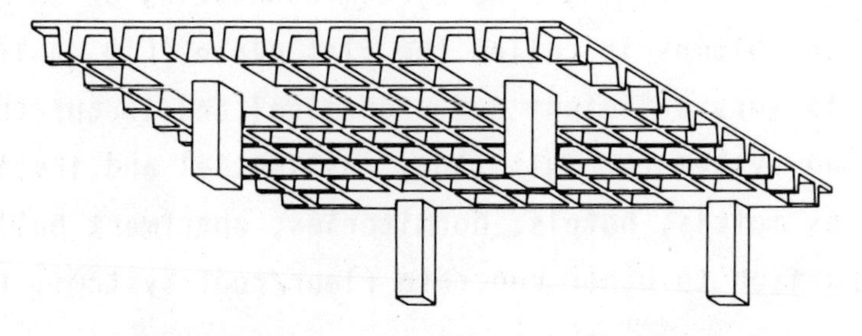

(d) Waffle Slab (Two-Way Joist Slab)

Fig. 4-1 Types of Two-Way Slab Systems

can be constructed in less time and with minimum labor costs because the system utilizes the simplest possible formwork and reinforcing steel layout. The use of flat plate construction also has other significant economic aspects. For instance, because of the shallow thickness of the floor system, story heights are automatically reduced resulting in smaller overall height of exterior walls and utility shafts; shorter floor to ceiling partitions; reductions in plumbing, sprinkler and duct risers; and a multitude of other items of construction. In cities like Washington, D.C., where the maximum height of buildings is restricted, the thin flat plate permits the construction of the maximum number of stories on a given plan area. Flat plates also provide for the most flexibility in the layout of columns, partitions, small openings, etc. Where job conditions allow direct application of the ceiling finish to the flat plate soffit, eliminating the need for suspended ceilings, additional cost and construction time savings are possible as compared to other structural systems.

The principal limitation on the use of flat plate construction is imposed by shear around the columns (Section 4.4). For heavy loads or long spans, the flat plate is often thickened locally around the columns creating what are known as drop panels. When a flat plate is equipped with drop panels, it is called a flat slab [Fig. 4-1(c)]. Also for reasons of shear around the columns, the column tops are sometimes flared, creating column capitals. For purposes of design, a column capital is part of the column, whereas a drop panel is part of the slab.

Waffle slab construction [Fig. 4-1(d)] consists of rows of concrete joists at right angles to each other with solid heads at the columns (for reasons of shear). The joists are commonly formed by using standard square "dome" forms. The domes are omitted around the columns to form the solid heads. For design purposes, waffle slabs are considered as flat slabs with the solid heads acting as drop panels. Waffle slab construction allows a considerable reduction in dead load as compared to conventional flat slab construction. Thus it is particularly advantageous where the use of long span and/or heavy loads is desired without the use of deepened drop panels or support beams.

The geometric shape formed by the joist ribs is often architecturally desirable.

Discussion in this chapter is limited largely to flat plates and flat slabs subjected only to gravity loads.

4.2 DEFLECTION CONTROL - MINIMUM SLAB THICKNESS

Minimum thickness/span ratios enable the designer to avoid extremely complex deflection calculations in routine designs. Deflections of two-way slab systems need not be computed if the overall slab thickness meets the minimum requirements specified in ACI 9.5.3. Minimum slab thicknesses for flat plates, flat slabs (and waffle slabs), and two-way slabs based on ACI Eqs. (9-11), (9-12), and (9-13) are summarized in Table 4-1, where ℓ_n is the clear span length in the long direction of a two-way slab panel. The tabulated values are the controlling minimum thicknesses governed by interior, side, or corner panels assuming a constant slab thickness for all panels making up a slab system.[4.1] Practical spandrel beam sizes will usually provide beam-to-slab stiffness ratios α greater than the minimum specified value of 0.8, so that a 10% increase in slab thickness for flat plate and flat slab panels with discontinuous edges would not be required. A "Standard" size drop panel that would allow a 10% reduction in the minimum required thickness of a flat slab floor system is illustrated in Fig. 4-2. Note: A larger size and depth drop may be used if required for shear strength; however, a corresponding lesser slab thickness is not permitted unless deflections are computed. The values for two-way slabs are based on an average beam-to-slab stiffness ratio α_m = 2.0; a lesser slab thickness may be used with higher beam-to-slab stiffness ratios per ACI Eq. (9-11).

For design convenience, minimum thicknesses for the six types of two-way slab systems listed in Table 4-1 are plotted in Fig. 4-3.

4.3 TWO-WAY SLAB ANALYSIS BY COEFFICIENTS

For gravity loads, ACI Chapter 13 provides two analysis methods for two-way slab systems, either the Direct Design Method or the Equivalent Frame Method. The Equivalent Frame Method, using member stiffnesses and involved analytical procedures, is not suitable for hand calculations. · Only the Direct Design Method, using moment coefficients, will be presented in this Chapter.

Table 4-1 - Minimum Thickness for Two-Way Slab Systems

Two-Way Slab System		Minimum h
Flat Plate	[Min. h = 5 in.]	$\ell_n/30$
Flat Plate with Spandrel Beams[1]		$\ell_n/33$
Flat Slab[2]	[Min. h = 4 in.]	$\ell_n/33$
Flat Slab[2] with Spandrel Beams[1]		$\ell_n/36$
Two-Way Slab[3] (square panels)	[Min. h = 3 1/2 in.]	$\ell_n/40$
Two-Way Slab[3] (rectangular 2:1 panels)		$\ell_n/46$

[1]Spandrel beam-to-slab stiffness ratio $\alpha \geq 0.8$ (ACI 9.5.3.3)
[2]Drop panel length $\geq \ell/3$, depth ≥ 1.25 h (ACI 9.5.3.2)
[3]Average beam-to-slab stiffness ratio $\alpha_m = 2.0$ [ACI 9.5.3.1(c)]

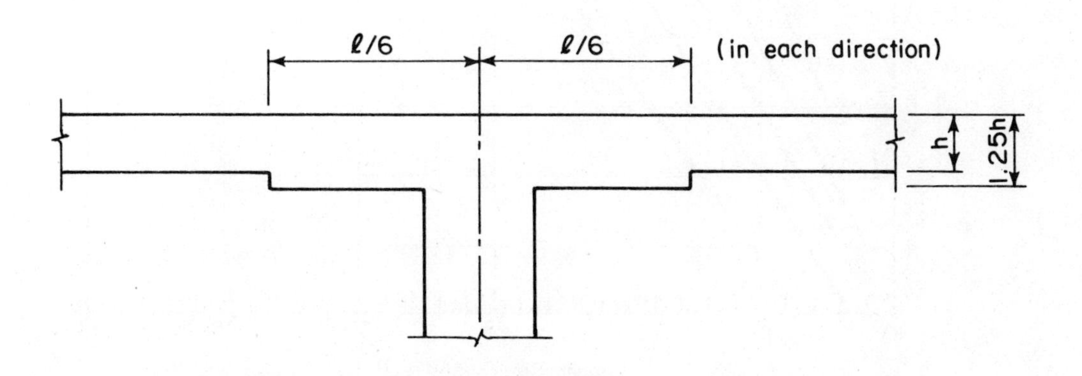

Fig. 4-2 Drop Panel Details (ACI 9.5.3.2)

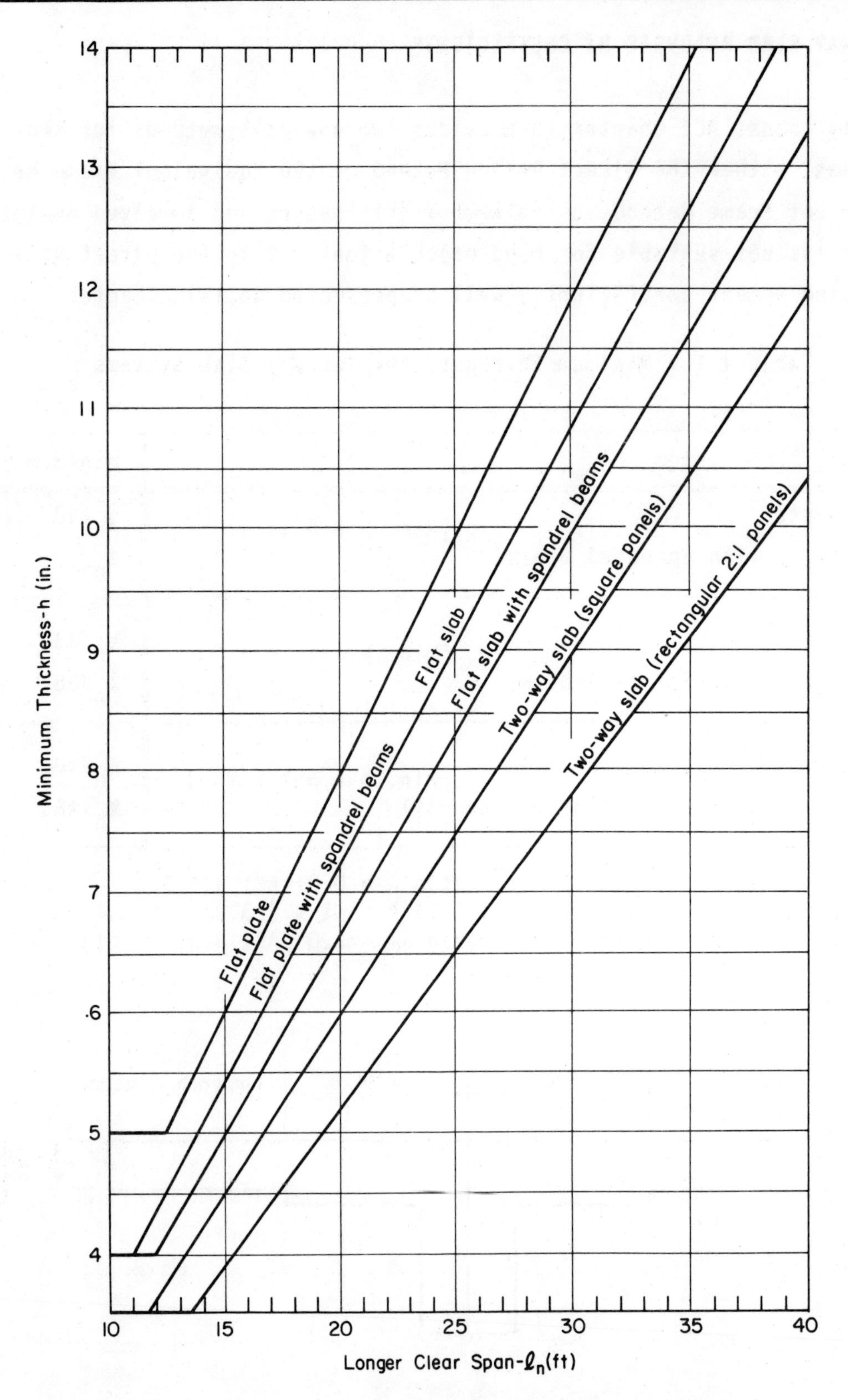

Fig. 4-3 Minimum Slab Thickness for Two-Way Slab Systems (See Table 4-1)

The Direct Design Method applies within the conditions illustrated in Fig. 4-4:

- There must be three or more continuous spans in each direction;

- Slab panels must be rectangular with a ratio of longer to shorter span (c/c of supports) not greater than 2;

- Successive span lengths (c/c of supports) in each direction must not differ by more than 1/3 of the longer span;

- Columns must not be offset more than 10% of the span (in direction of offset) from either axis between centerlines of successive columns;

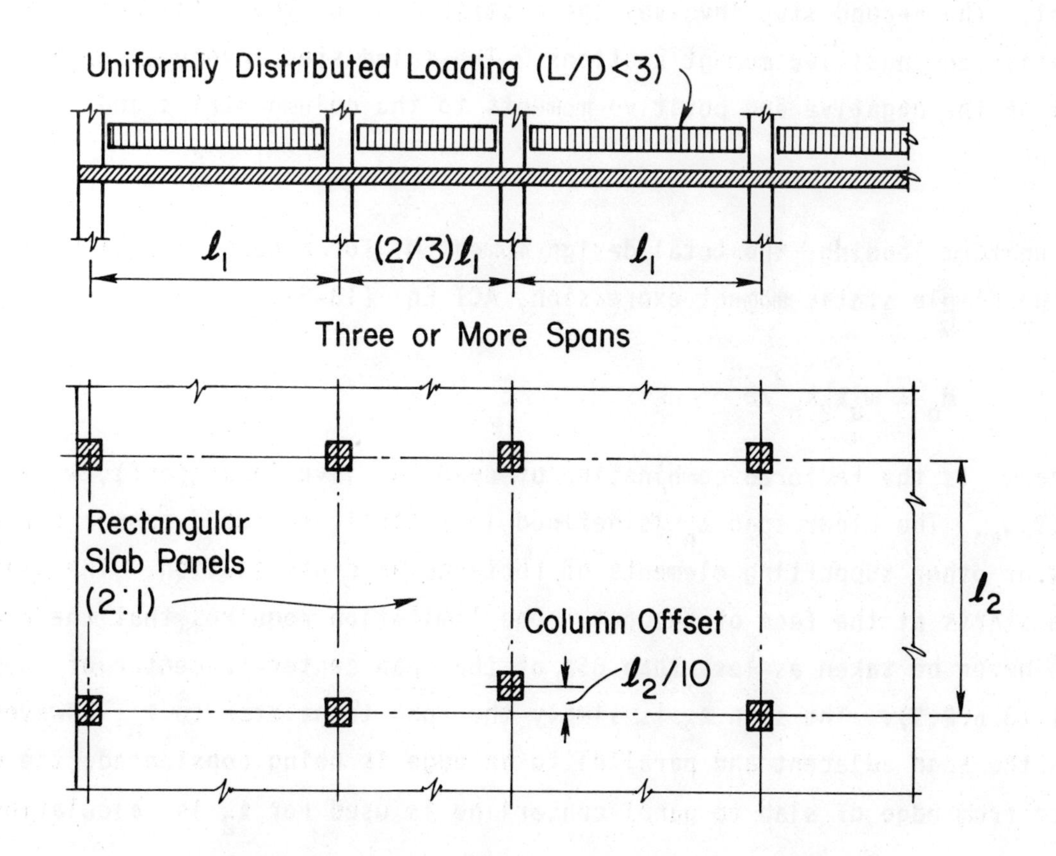

Fig. 4-4 – Conditions for Analysis by Coefficients

- Loads must be uniformly distributed with the live load not more than 3 times the dead ($L/D \leq 3$). Note that if the live load exceeds one-half the dead ($L/D > 1/2$), column-to-slab stiffness ratios must exceed certain values given in ACI Table 13.6.10, or positive factored moments in panels supported by columns not meeting such minimum stiffness requirements must be magnified by a coefficient computed by ACI Eq. (13-5);

- For two-way slabs, relative stiffness of beams in two perpendicular directions must satisfy the minimum and maximum requirements given in ACI 13.6.1.6;

- Redistribution of moments by ACI 8.4 is not permitted.

In essence, the Direct Design Method is a three-step analysis procedure. The first step is the calculation of the total design moment M_o for a given panel. The second step involves the distribution of the total moment to the negative and positive moment sections. The third step involves the assignment of the negative and positive moments to the column strips and middle strips.

For uniform loading, the total design moment M_o for a panel is calculated by the simple static moment expression, ACI Eq. (13-3):

$$M_o = w_u \ell_2 \ell_n^2/8$$

where w_u is the factored combination of dead and live loads (psf), $w_u = 1.4\ w_d + 1.7\ w_\ell$. The clear span ℓ_n is defined in a straightforward manner for columns or other supporting elements of rectangular cross section. The clear span starts at the face of support. One limitation requires that the clear span never be taken as less than 65% of the span center-to-center of supports (ACI 13.6.2.5). The span ℓ_2 is simply the span transverse to ℓ_n; however, when the span adjacent and parallel to an edge is being considered, the distance from edge of slab to panel centerline is used for ℓ_2 in calculation of M_o.

Division of the total panel moment M_o into negative and positive moments, and then, column and middle strip moments, involves direct application of moment coefficients to the total moment M_o. The moment coefficients are a function of span (interior or exterior) and slab support conditions (type of two-way slab system). For design convenience, moment coefficients for typical two-way slab systems are given in Tables 4-2 through 4-6. Tables 4-2 through 4-5 apply to flat plates or flat slabs with differing end support conditions. Table 4-6 applies to two-way slabs (slabs supported on beams on all four sides). Final moments for the column strip and middle strip are computed directly using the tabulated values.

The moment coefficients of Table 4-3 (flat plate with spandrel beams) are valid for $\beta_t \geq 2.5$. The coefficients of Table 4-6 (two-way slabs) apply for $\alpha_1 \ell_2/\ell_1 \geq 1.0$ and $\beta_t \geq 2.5$. Many practical beam sizes will provide beam-to-slab stiffness ratios such that $\alpha_1 \ell_2/\ell_1$ and β_t would be greater than these limits allowing moment coefficients to be taken directly from the Tables without further consideration of stiffnesses and interpolation for moment coefficients. However, if beams are present, the two stiffness parameters α_1 and β_t will need to be evaluated. For two-way slabs, the stiffness parameter α_1 is simply the ratio of the moments of inertia of the effective beam and slab sections in the direction of analysis, $\alpha_1 = I_b/I_s$, as illustrated in Fig. 4-5. Figures 4-6 and 4-7 simplify evaluation of the α term.

Relative stiffness provided by a spandrel beam is reflected by the parameters $\beta_t = C/2I_s$, where I_s is the moment of inertia of the effective slab section spanning in the direction of ℓ_1 and having a width equal to ℓ_2, $I_s = \ell_2 h^3/12$. The constant C pertains to the torsional stiffness of the effective spandrel beam cross section. It is found by dividing the beam section into its component rectangles, each having smaller dimension x and larger dimension y, and summing the contribution of all the parts by means of the equation

$$C = \Sigma(1 - 0.63\frac{x}{y})\frac{x^3 y}{3}$$

The subdivision can be done in such a way as to maximize C. Figure 4-8 simplifies calculation of torsional constant C.

Table 4-2 - Flat Plate or Flat Slab Supported Directly on Columns

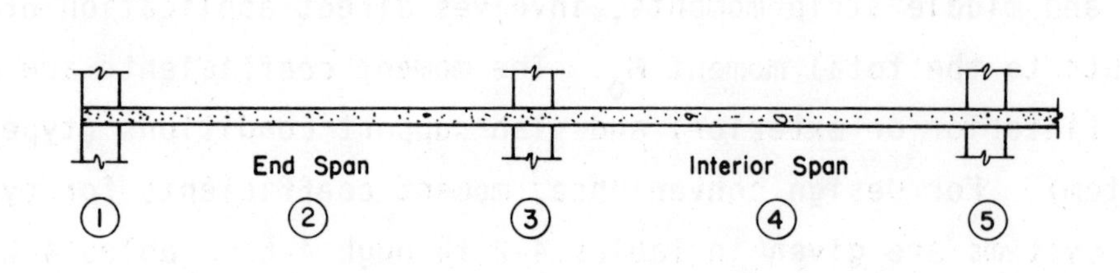

Slab Moments	End Span			Interior Span	
	① Exterior Negative	② Positive	③ First Interior Negative	④ Positive	⑤ Interior Negative
Total Moment	$0.26\ M_o$	$0.52\ M_o$	$0.70\ M_o$	$0.35\ M_o$	$0.65\ M_o$
Column Strip	$0.26\ M_o$	$0.31\ M_o$	$0.53\ M_o$	$0.21\ M_o$	$0.49\ M_o$
Middle Strip	0	$0.21\ M_o$	$0.17\ M_o$	$0.14\ M_o$	$0.16\ M_o$

Notes: (1) All negative moments are at face of support

Table 4-3 - Flat Plate or Flat Slab with Spandrel Beams

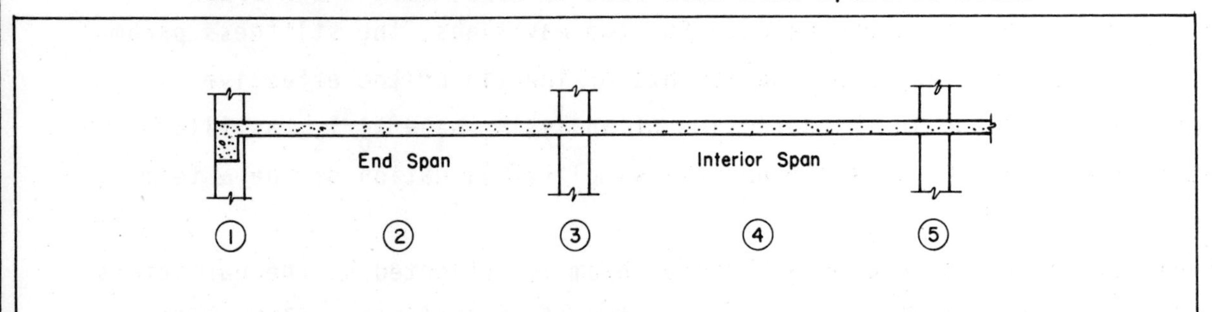

Slab Moments	End Span			Interior Span	
	① Exterior Negative	② Positive	③ First Interior Negative	④ Positive	⑤ Interior Negative
Total Moment	$0.30\ M_o$	$0.50\ M_o$	$0.70\ M_o$	$0.35\ M_o$	$0.65\ M_o$
Column Strip	$0.23\ M_o$	$0.30\ M_o$	$0.53\ M_o$	$0.21\ M_o$	$0.49\ M_o$
Middle Strip	$0.07\ M_o$	$0.20\ M_o$	$0.17\ M_o$	$0.14\ M_o$	$0.16\ M_o$

Notes: (1) All negative moments are at face of support
(2) Torsional stiffness of spandrel beams $\beta_t \geq 2.5$. For values of β_t less than 2.5, exterior negative column strip moment increases to $(0.30 - .03\ \beta_t)\ M_o$.

Table 4-4 - Flat Plate or Flat Slab with End Span Integral with Wall

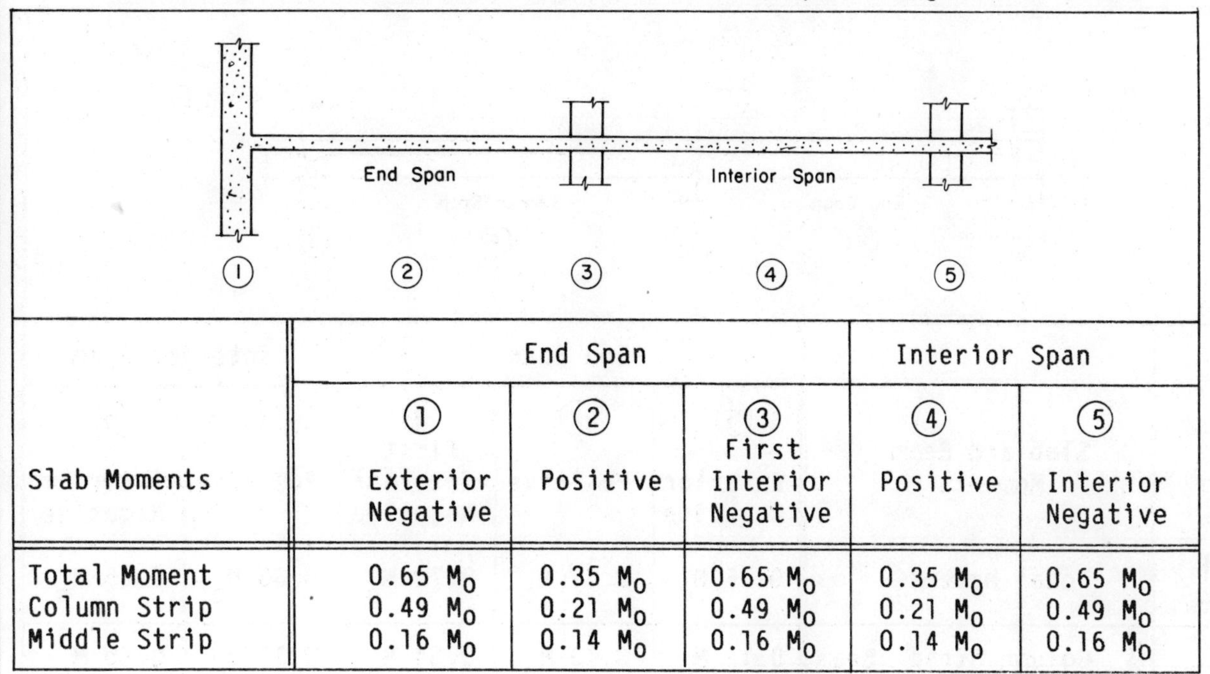

Slab Moments	End Span ① Exterior Negative	② Positive	③ First Interior Negative	Interior Span ④ Positive	⑤ Interior Negative
Total Moment	0.65 M_o	0.35 M_o	0.65 M_o	0.35 M_o	0.65 M_o
Column Strip	0.49 M_o	0.21 M_o	0.49 M_o	0.21 M_o	0.49 M_o
Middle Strip	0.16 M_o	0.14 M_o	0.16 M_o	0.14 M_o	0.16 M_o

Notes: (1) All negative moments are at face of support

Table 4-5 - Flat Plate or Flat Slab with End Span Simply Supported on Wall

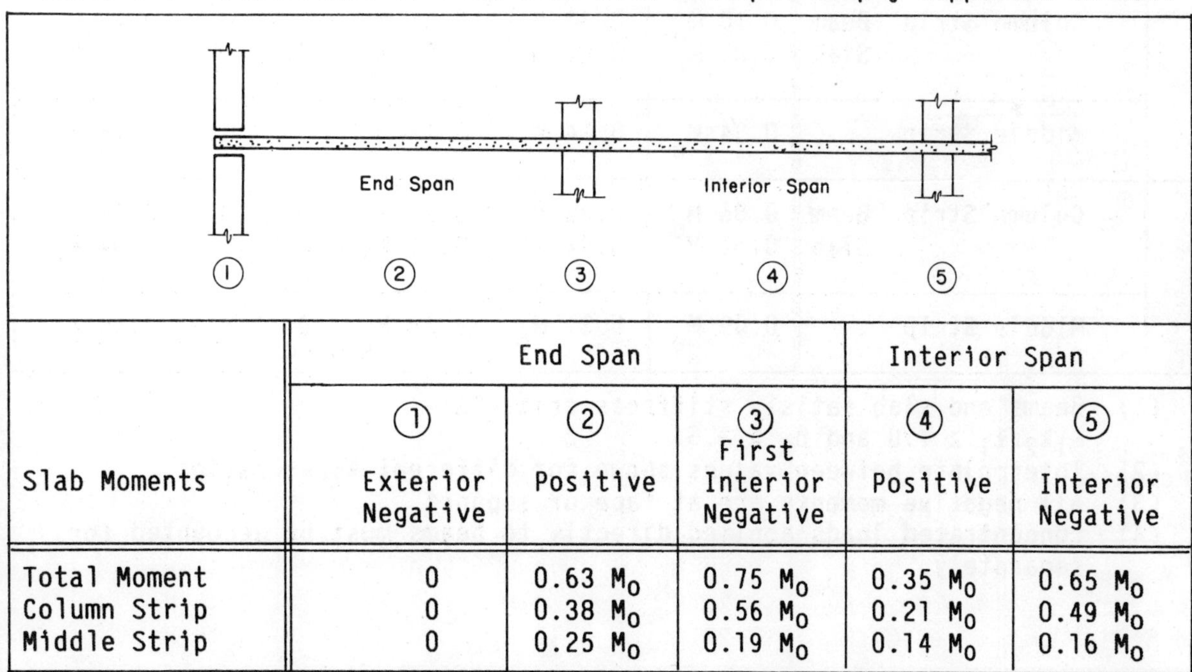

Slab Moments	End Span ① Exterior Negative	② Positive	③ First Interior Negative	Interior Span ④ Positive	⑤ Interior Negative
Total Moment	0	0.63 M_o	0.75 M_o	0.35 M_o	0.65 M_o
Column Strip	0	0.38 M_o	0.56 M_o	0.21 M_o	0.49 M_o
Middle Strip	0	0.25 M_o	0.19 M_o	0.14 M_o	0.16 M_o

Notes: (1) All negative moments are at face of support

Table 4-6 - Two-Way Slab

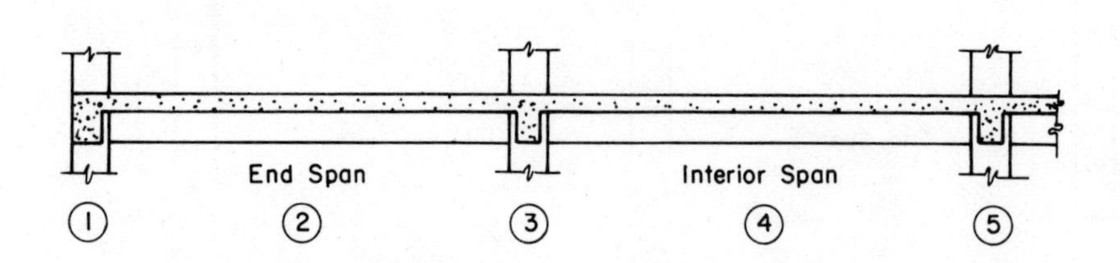

Span ratio ℓ_2/ℓ_1	Slab and Beam Moments		End Span			Interior Span	
			① Exterior Negative	② Positive	③ First Interior Negative	④ Positive	⑤ Interior Negative
	Total Moment		$0.16\ M_o$	$0.57\ M_o$	$0.70\ M_o$	$0.35\ M_o$	$0.65\ M_o$
0.5	Column Strip	Beam	$0.12\ M_o$	$0.43\ M_o$	$0.54\ M_o$	$0.27\ M_o$	$0.50\ M_o$
		Slab	$0.02\ M_o$	$0.08\ M_o$	$0.09\ M_o$	$0.05\ M_o$	$0.09\ M_o$
	Middle Strip		$0.02\ M_o$	$0.06\ M_o$	$0.07\ M_o$	$0.03\ M_o$	$0.06\ M_o$
1.0	Column Strip	Beam	$0.10\ M_o$	$0.37\ M_o$	$0.45\ M_o$	$0.22\ M_o$	$0.42\ M_o$
		Slab	$0.02\ M_o$	$0.06\ M_o$	$0.08\ M_o$	$0.04\ M_o$	$0.07\ M_o$
	Middle Strip		$0.04\ M_o$	$0.14\ M_o$	$0.17\ M_o$	$0.09\ M_o$	$0.16\ M_o$
2.0	Column Strip	Beam	$0.06\ M_o$	$0.22\ M_o$	$0.27\ M_o$	$0.14\ M_o$	$0.25\ M_o$
		Slab	$0.01\ M_o$	$0.04\ M_o$	$0.05\ M_o$	$0.02\ M_o$	$0.04\ M_o$
	Middle Strip		$0.09\ M_o$	$0.31\ M_o$	$0.38\ M_o$	$0.19\ M_o$	$0.36\ M_o$

Notes:
(1) Beams and slab satisfy stiffness criteria: $\alpha_1\ell_2/\ell_1 \geq 1.0$ and $\beta_t \geq 2.5$.
(2) Interpolate between values shown for different ℓ_2/ℓ_1 ratios.
(3) All negative moments are at face of support.
(4) Concentrated loads applied directly to beams must be accounted for separately.

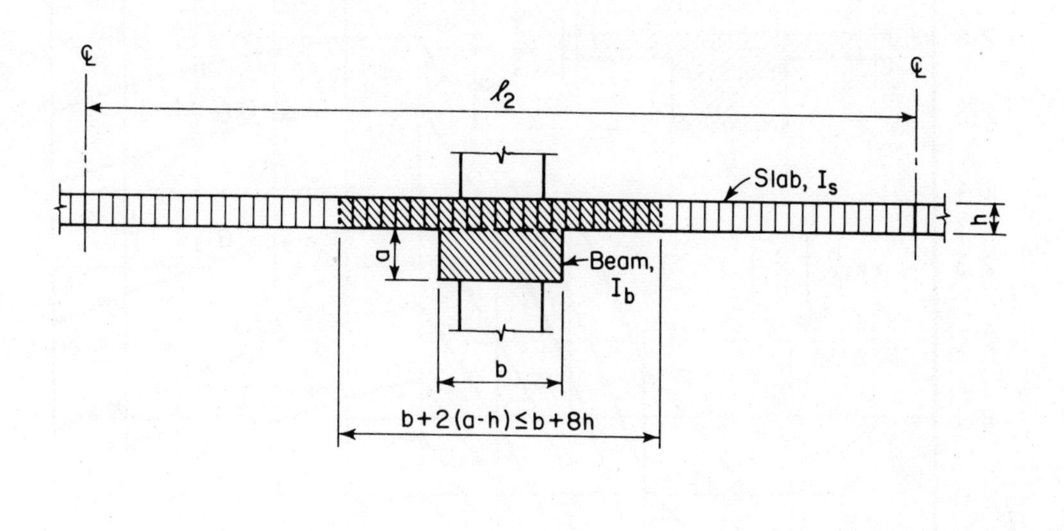

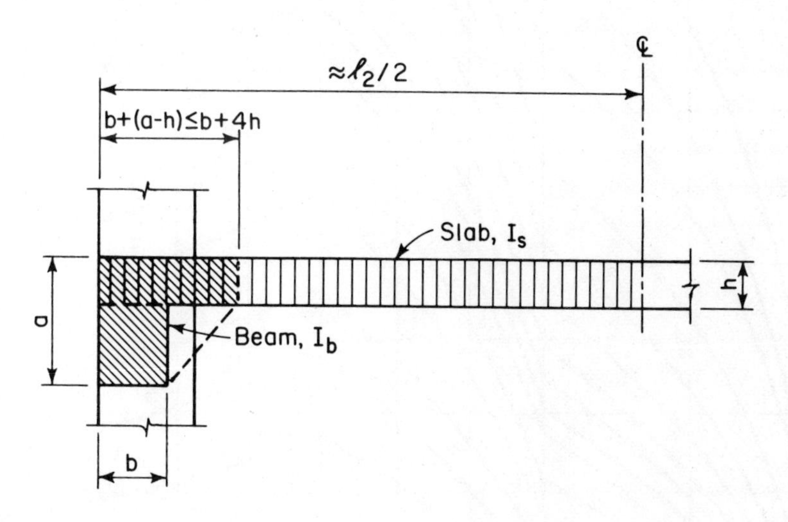

Fig. 4-5 Effective Beam and Slab Sections for Stiffness Ratio α

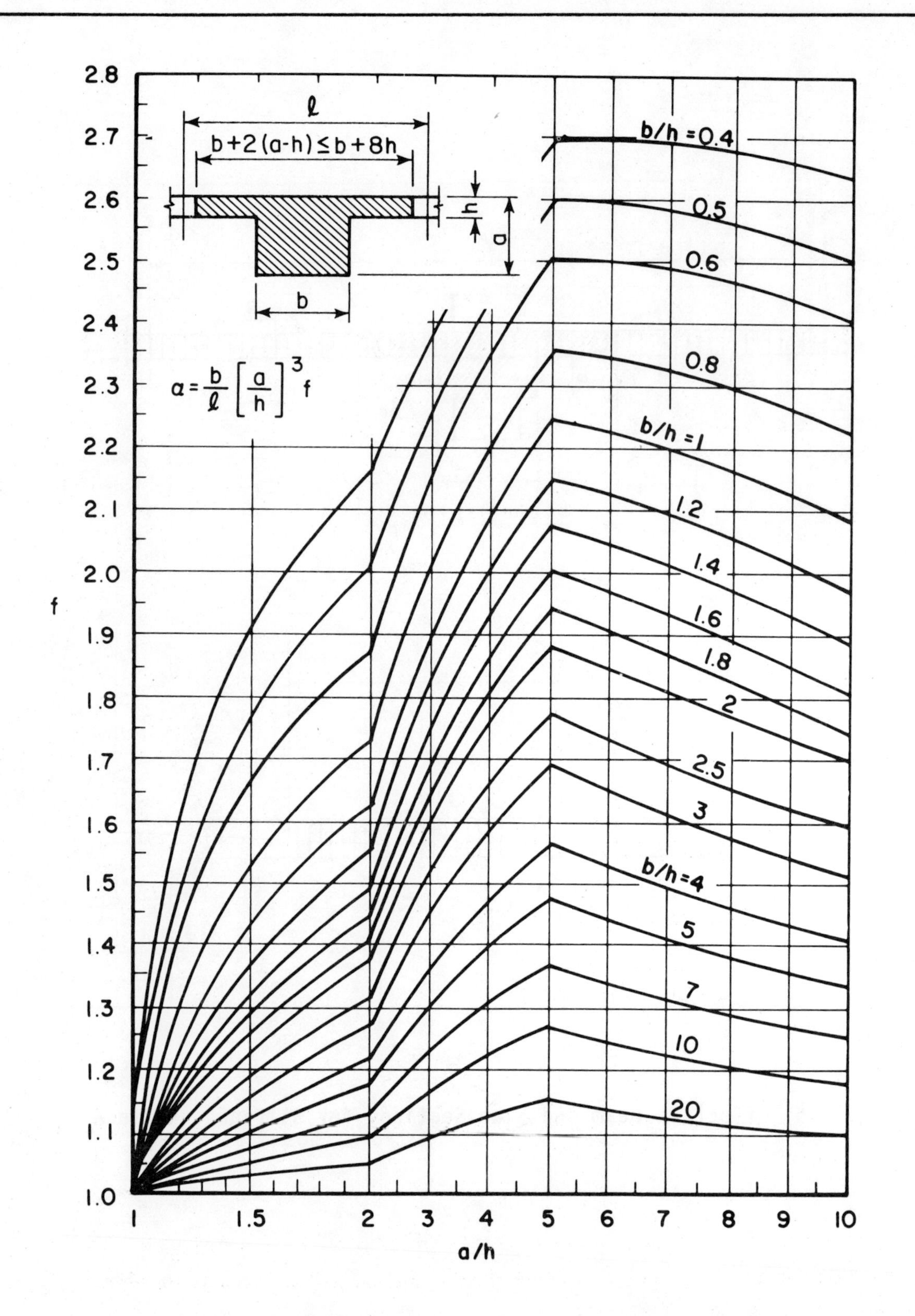

Fig. 4-6 Beam to Slab Stiffness Ratio α (Interior Beams)

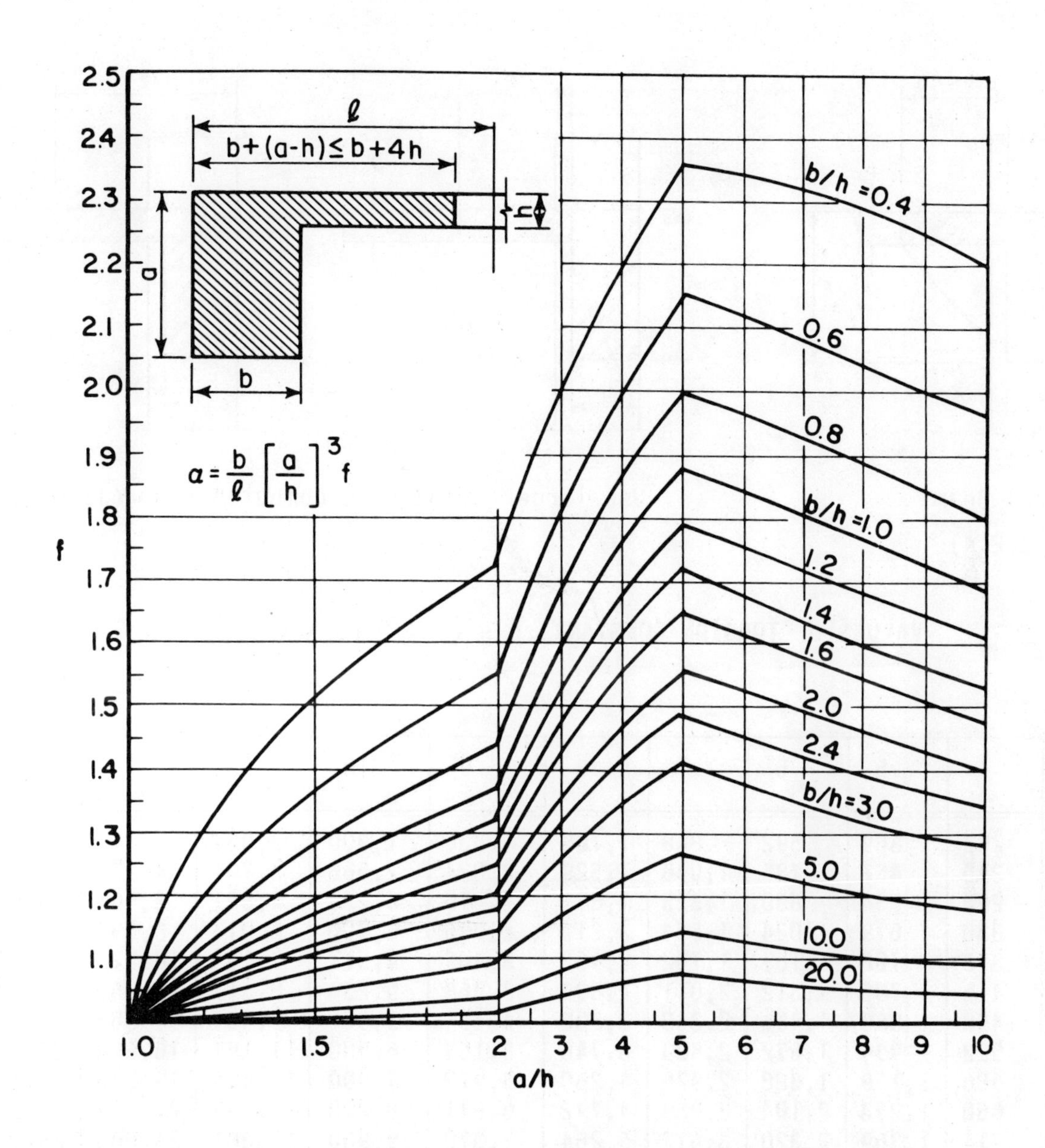

Fig. 4-7 Beam to Slab Stiffness Ratio α (Spandrel Beams)

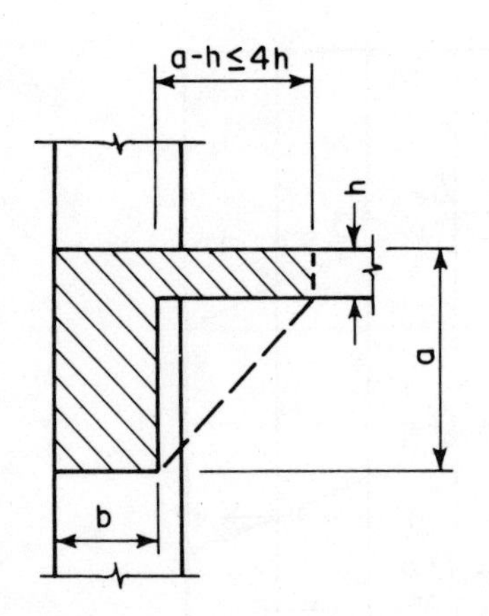

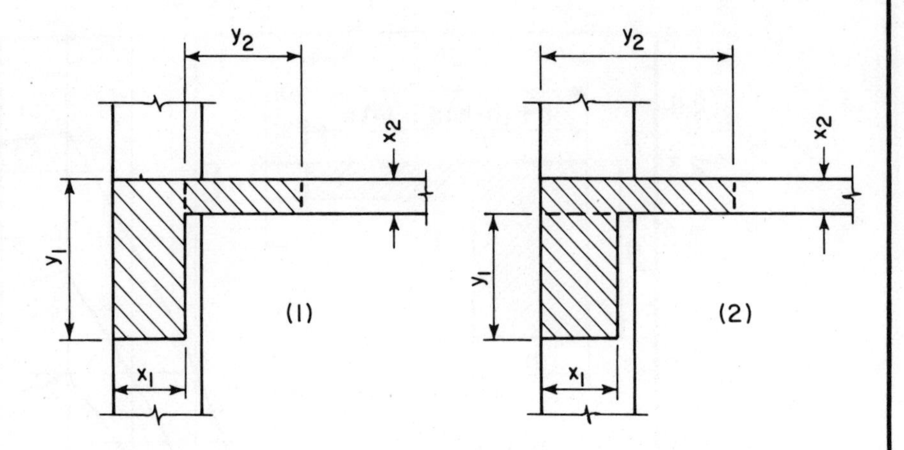

Spandrel Beam
(ACI 13.2.4)

Use larger value of C computed from (1) or (2)

VALUES OF TORSION CONSTANT, $C = (1 - 0.63\ x/y)(x^3y/3)$

y \ x*	4	5	6	7	8	9	10	12	14	16
12	202	369	592	868	1,188	1,538	1,900	2,557	--	--
14	245	452	736	1,096	1,529	2,024	2,566	3,709	4,738	--
16	288	534	880	1,325	1,871	2,510	3,233	4,861	6,567	8,083
18	330	619	1,024	1,554	2,212	2,996	3,900	6,013	8,397	10,813
20	373	702	1,167	1,782	2,553	3,482	4,567	7,165	10,226	13,544
22	416	785	1,312	2,011	2,895	3,968	5,233	8,317	12,055	16,275
24	458	869	1,456	2,240	3,236	4,454	5,900	9,469	13,885	19,005
27	522	994	1,672	2,583	3,748	5,183	6,900	11,197	16,628	23,101
30	586	1,119	1,888	2,926	4,260	5,912	7,900	12,925	19,373	27,197
33	650	1,243	2,104	3,269	4,772	6,641	8,900	14,653	22,117	31,293
36	714	1,369	2,320	3,612	5,284	7,370	9,900	16,381	24,860	35,389
42	842	1,619	2,752	4,298	6,308	8,828	11,900	19,837	30,349	43,581
48	970	1,869	3,183	4,984	7,332	10,286	13,900	23,293	35,836	51,773
54	1,098	2,119	3,616	5,670	8,356	11,744	15,900	26,749	41,325	59,965
60	1,226	2,369	4,048	6,356	9,380	13,202	17,900	30,205	46,813	68,157

*Small side of a rectangular cross section with dimensions x and y.

Fig. 4-8 - Design Aid for Computing Torsional Section Constant C

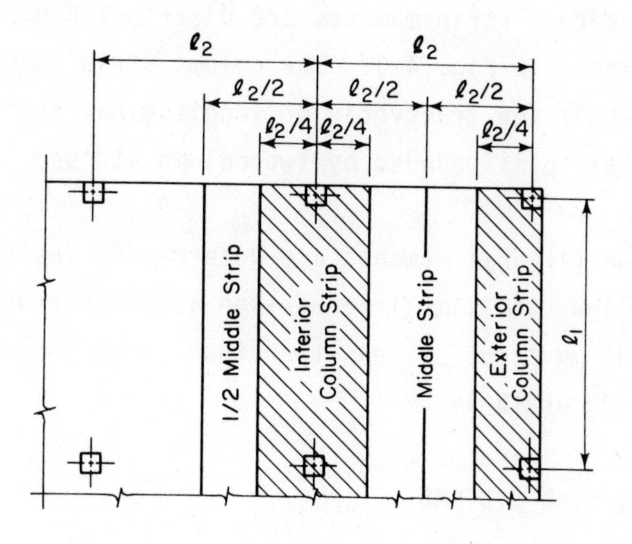

(a) Column Strip for $\ell_2 \leq \ell_1$

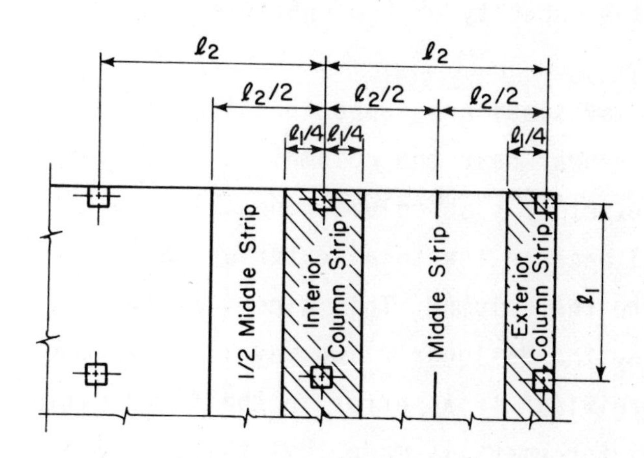

(b) Column Strip for $\ell_2 > \ell_1$

Fig. 4-9 Definition of Design Strips

The column strip and middle strip moments are distributed over an effective slab width as illustrated in Fig. 4-9. The column strip is defined as having a width equal to one-half the transverse or longitudinal span, whichever is smaller. The middle strip is bounded by two column strips.

Once the slab and beam (if any) moments are determined, design of the slab and beam sections follows the simplified design approach presented in Chapter 3. Slab reinforcement must not be less than that given in Table 3-5...with a maximum spacing of 2h or 18 in.

4.4 SHEAR IN TWO-WAY SLAB SYSTEMS

If two-way slab systems are supported by beams or walls, the slab shear is seldom a critical factor in design, as the shear force at factored loads is generally well below the capacity of the concrete.

In contrast, when two-way slabs are supported directly by columns as in flat plates and flat slabs, shear near the columns is of critical importance. Shear strength at an exterior slab-column connection (without spandrel beams) is especially critical because the total exterior negative slab moment must be transferred directly to the column. This aspect of two-way slab design should not be taken lightly by the designer. Two-way slab systems will normally be found to be quite "forgiving" if an error in the distribution or even in the amount of flexural reinforcement is made, but there will be no forgiveness if a critical lapse occurs in providing the required shear strength.

For slab systems supported directly by columns, it is advisable at an early stage in the design to check the shear strength of the slab in the vicinity of columns as illustrated in Fig. 4-10.

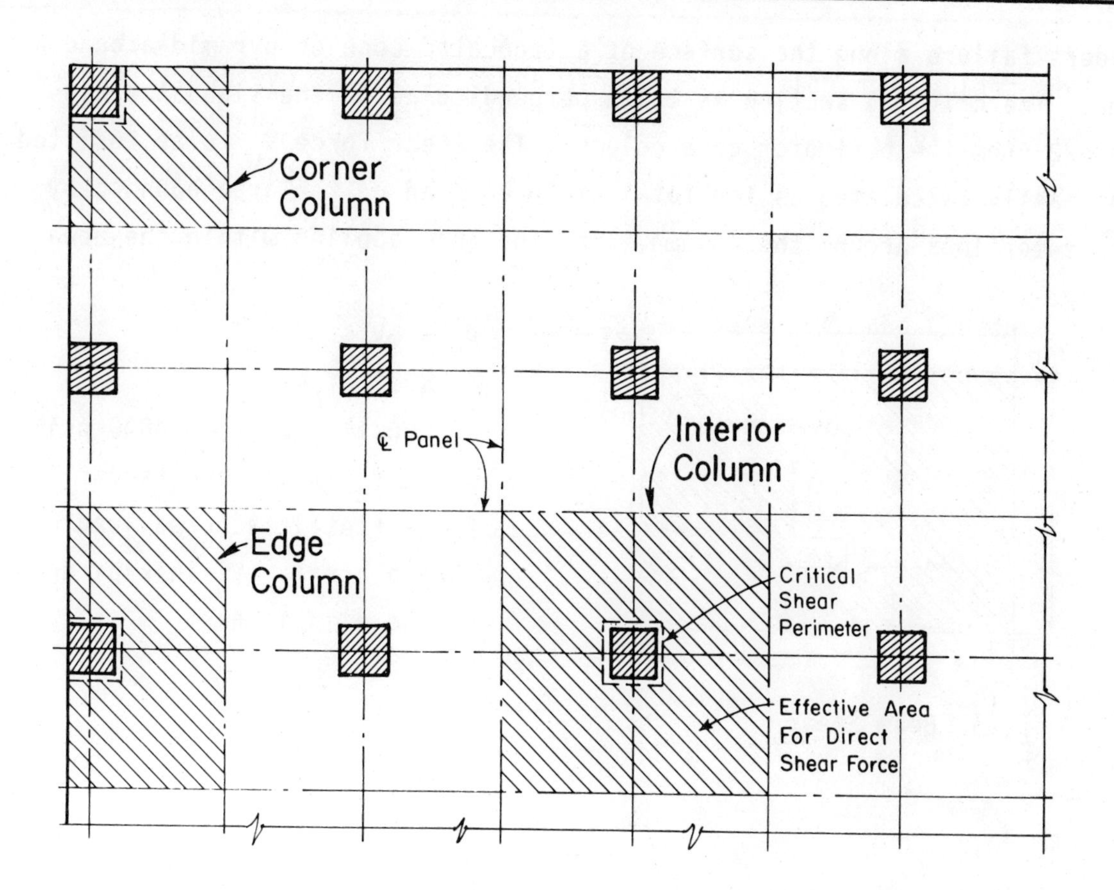

Fig. 4-10 Critical Locations for Slab Shear Strength

4.4.1 Shear in Flat Plate and Flat Slab Floor Systems

Two types of shear need to be considered in the design of flat plates or flat slabs supported directly on columns. The first is the familiar one-way or beam-type shear, which may be critical in long narrow slabs. Analysis for beam shear considers the slab to act as a wide beam spanning between the columns. The critical section is taken a distance d from the face of the column. Design against beam shear consists in checking for satisfaction of the requirement indicated in Fig. 4-11(a). Beam shear in slabs is seldom a critical factor in design, as the shear force is usually well below the shear capacity of the concrete.

Two-way or "punching" shear is generally the more critical of the two types of shear in slab systems supported directly on columns. Punching shear considers failure along the surface of a truncated cone or pyramid around a column. The critical section is taken perpendicular to the slab at a distance d/2 from the perimeter of a column. The shear force V_u to be resisted can be easily calculated as the total factored load on the area bounded by panel centerlines around the column, less the load applied within the area

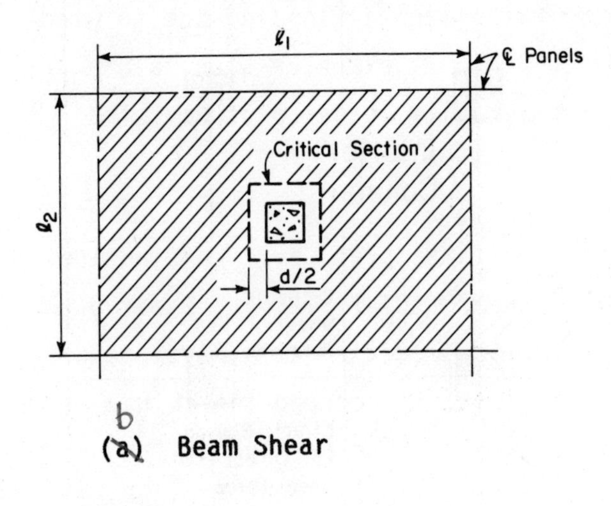

(a) Beam Shear

$V_u \leq \varphi V_c$

$\leq \varphi 2 \sqrt{f'_c} \ell_2 d$

$\leq 0.11\ \ell_2 d$ (f'_c = 4000 psi)

where V_u is factored shear force (total factored load on shaded area). V_u in kips and ℓ_2 and d in inches.

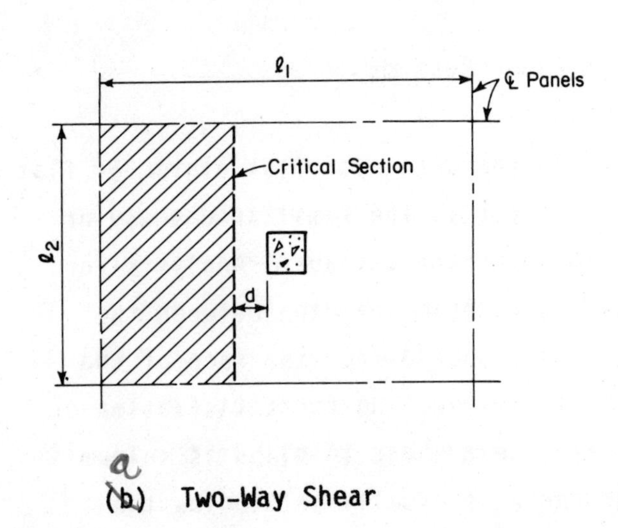

(b) Two-Way Shear

$V_u \leq \varphi V_c$

$\leq \varphi 4 \sqrt{f'_c} b_o d$ (square columns)

$\leq 0.22\ b_o d$ (f'_c = 4000 psi)

where V_u is factored shear force (total factored load on shaded area) and b_o is perimeter of critical section. See Table 4-7 for values of A_c = $b_o d$. V_u in kips and b_o and d in inches.

Fig. 4-11 Direct Shear at an Interior Slab-Column Support. See Fig. 4-10

defined by the critical shear perimeter. See Fig. 4-10. In the absence of a significant moment transfer from the slab to the column, design against punching shear consists in making sure that the requirement of Fig. 4-11(b) is satisfied. For practical design, only direct shear (uniformly distributed around the perimeter b_o) occurs around interior slab-column supports where no (or insignificant) moment is to be transferred from the slab to the column. Significant moments may have to be carried when unbalanced gravity loads on either side of an interior column or horizontal loading due to wind must be transferred from the slab to the column. At exterior slab-column supports the total exterior slab moment from gravity loads (plus any wind moments) must be transferred directly to the column.

Transfer of moment between a slab and a column takes place by a combination of flexure (ACI 13.3.3) and eccentricity of shear (ACI 11.12.2). Shear due to moment transfer is assumed to act on a critical section at a distance d/2 from the face of the column, the same critical section around the column as that used for direct shear transfer [Fig. 4-11(b)]. The portion of the moment transferred by flexure is assumed to be transferred over a width of slab equal to the transverse column width c_2, plus 1.5 times the slab thickness (1.5h) on either side of the column. Concentration of negative reinforcement is to be used to resist moment on this effective slab width. The combined shear stress due to direct shear and moment transfer often governs the design, especially at the exterior slab-column supports.

The portions of the total moment to be transferred by eccentricity of shear and by flexure are given by ACI Eqs. (11-40) and (13-1). For square columns, 40% of the moment is considered transferred by eccentricity of the shear ($\gamma_v M_u = 0.40 M_u$), and 60% by flexure ($\gamma_f M_u = 0.60 M_u$), where M_u is the transfer moment at the centroid of the critical section. The moment M_u at an exterior slab-column support will generally not be computed at the centroid of the critical transfer section in the frame analysis. In the Direct Design Method moments are computed at the face of support. Considering the approximate nature of the procedure used to evaluate the stress distribution due to moment transfer, it seems unwarranted to consider a change in moment to the transfer centroid...use of the moment values at face of support directly would usually be accurate enough.

The factored shear stress on the critical transfer section is the sum of the direct shear and shear caused by moment transfer,

$$v_u = V_u/A_c + \gamma_v M_u\, c/J$$

or

$$v_u = V_u/A_c - \gamma_v M_u\, c'/J$$

For slabs supported on square columns, shear stress v_u must not exceed $\varphi 4\sqrt{f'_c}$. For f'_c = 4000 psi, $v_u \leq$ 215 psi.
Computation of the combined shear stress involves the following properties of the critical transfer section.

A_c = area of critical section

c or c' = distance from centroid of critical section to face of section where stress is being computed

J_c = property of critical section analogous to polar moment of inertia

The above properties are given in terms of formulas in Tables 4-7 through 4-10 for the four cases that can arise with a rectangular column section: (Table 4-7) interior column, (Table 4-8) edge column with bending parallel to the edge, (Table 4-9) edge column with bending perpendicular to the edge, and (Table 4-10) corner column. Numerical values of the above parameters for various combinations of square column sizes and slab thicknesses are also listed in Tables 4-7 through 4-10. Note that in the case of flat slabs, two different critical sections need to be considered in punching shear calculations as shown in Fig. 4-12. Tables 4-7 through 4-10 can be used in both cases.

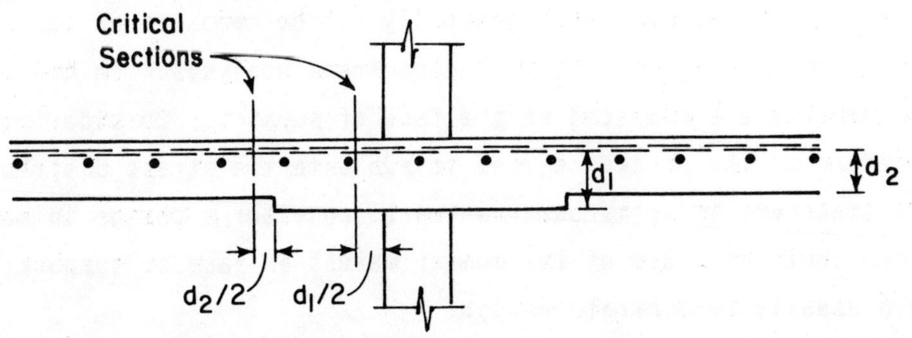

Fig. 4-12 Critical Shear-Transfer Sections for Flat Slabs

Unbalanced moment transfer between slab and an edge column (without spandrel beams) requires special consideration when slabs are analyzed by the Direct Design Method for gravity load. To assure adequate shear strength when using the approximate end-span moment coefficient, the full nominal moment strength M_n provided by the column strip must be used to calculate the portion of moment transferred by eccentricity of shear ($\gamma_v M_u = 0.40\ M_n$ of column strip) according to ACI 13.6.3.6. For end spans without spandrel beams, the column strip is proportioned to resist the total exterior negative factored moment (Table 4-2). The above requirement is illustrated in Fig. 4-13. The total reinforcement provided in the column strip includes the additional reinforcement concentrated over the column to resist the fraction of unbalanced moment transferred by flexure $\gamma_f M_u = 0.60\ M_u = 0.60(0.26 M_o)$, where the moment coefficient (0.26) is from Table 4-2. Application of this special design requirement is illustrated in Section 4.7 - Example: Two-way Flat Plate.

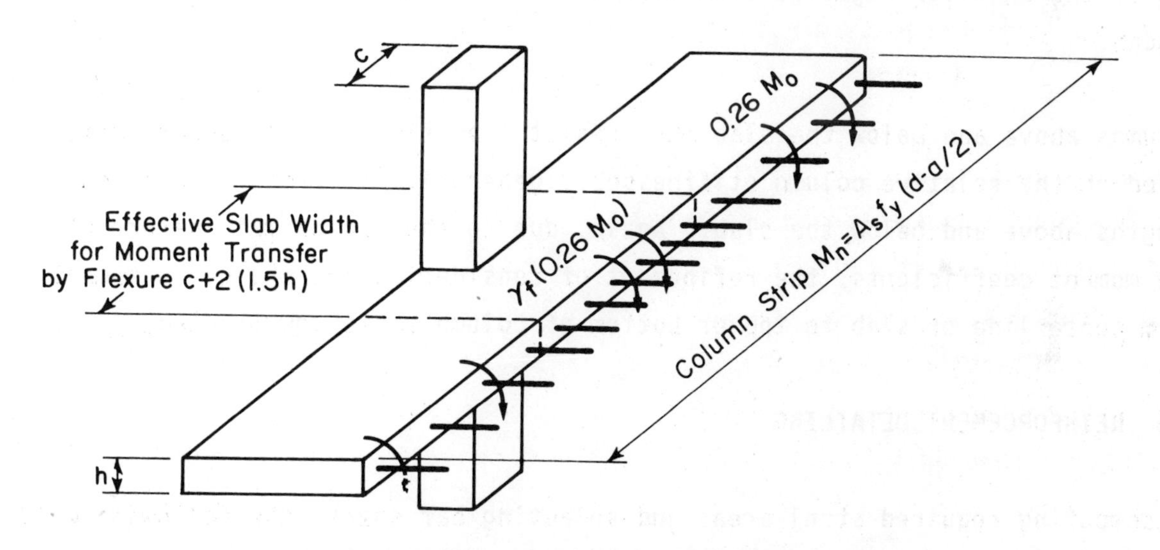

Fig. 4-13 Nominal Moment Strength of Column Strip for Evaluation of $\gamma_v M_u = 0.40\ M_n$

4.5 COLUMN MOMENTS DUE TO GRAVITY LOADS

Supporting columns (and walls) must resist any negative moments transferred from the slab system. For interior columns, the approximate ACI Eq. (13-4) may be used for unbalanced moment transfer due to gravity loading, unless an analysis is made considering effects of pattern loading and unequal adjacent spans. The transfer moment is computed directly as a function of the span length and gravity loading. For the more usual case with equal transverse and adjacent spans, ACI Eq. (13-4) simplifies to:

$$M_u = 0.07(0.5\ w_\ell \ell_2 \ell_n^{\ 2}) = 0.035\ w_\ell \ell_2 \ell_n^{\ 2}$$

where w_ℓ = factored live load, psf

ℓ_2 = span length transverse to ℓ_n

ℓ_n = clear span length in direction M_u is being determined

At an exterior column, the total exterior negative moment from the slab system is transferred directly to the column. Due to the approximate nature of the moment coefficients of the Direct Design Method, it seems unwarranted to consider the change in moment from face of support to centerline of support... use of the exterior negative slab moment directly would usually be accurate enough.

Columns above and below the slab must resist a portion of the support moment based on the relative column stiffnesses...generally, in proportion to column lengths above and below the slab. Again, due to the approximate nature of the moment coefficients, the refinement of considering the change in moment from centerline of slab to top or bottom of column seems unwarranted.

4.6 REINFORCEMENT DETAILING

In computing required steel areas and selecting bar sizes, the following will ensure conformance to the Code and a practical design.

(1) Minimum reinforcement area = 0.0018 bh (b = slab width, h = total thickness) for Grade 60 bars for either top or bottom steel. These minima apply separately in each direction.

(2) Maximum bar spacing is 2h, but not more than 18 in.

(3) Maximum top bar spacing at all interior locations subject to construction traffic should be limited. Not less than #4 @ 12 in. is recommended to provide adequate rigidity and to avoid displacement of top bars with standard bar support layouts under ordinary foot traffic.

(4) Maximum $\rho = A_s/bd$ is limited to 0.75 ρ_b (ρ_b = balanced reinforcement ratio); however, $\rho_{max} \leq 0.50\rho_b$ is recommended to provide deformability, to avoid overly flexible systems subject to objectionable vibration or deflection, and for a practical balance to achieve overall economy of materials, construction and design time.

(5) Generally, the largest size of bars that will satisfy the maximum limits on spacing will provide overall economy. Critical dimensions that limit size are the thickness of slab available for hooks and the distance from the critical design sections to edge of slab.

4.7 EXAMPLES: SIMPLIFIED DESIGN FOR TWO-WAY SLABS (see p 1-13)

The following two examples illustrate use of the simplified design data presented in Chapter 4 for analysis and design of two-way slab systems. The two-way slab system for Bldg. #2 is used to illustrate simplified design.

4.7.1 - Example: Design a typical interior strip (NS-direction) of the two-way flat plate of Alternate (2) framing. Slab and column framing designed for gravity load only; structural walls designed for total wind forces.

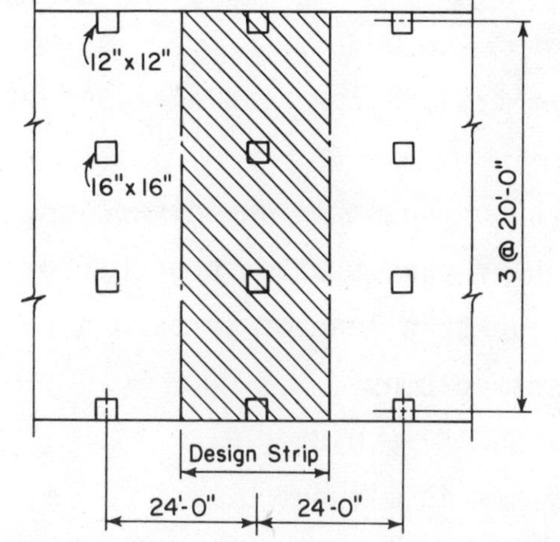

(1) Data: f'_c = 4000 psi
f_y = 60,000 psi

Floors: LL = 50 psf
DL = 142 psf (9 in. slab + 20 psf partitions + 10 psf ceiling & misc.) p1-12

Preliminary sizing: Slab (without spandrels) = 9 in.

Columns int. = 16x16 } see p5-24

ext. = 12x12

(2) Check slab thickness for deflection control and shear strength.

(a) Deflection control

From Table 4-1 (flat plate): $h = \ell_n/30 = (22.67 \times 12)/30 = 9.07$ in. ← always long direction

where $\ell_n = 24 - 16/12 = 22.67$ ft

(b) Shear strength

From Fig 4-11 (p4-20): Check two-way shear strength at interior slab-column support for h = 9 in.

From Table 4-7 (p4-45): $A_c = 720.6$ in.^{2} for 9 in. slab with 16x16 column.

$a = (11.81)2 = 23.62$ in $= 1.97$ ft.

or $C_1 + 2d/2 = 16 + 7\frac{5}{8} = 23.62''$

live load reduction - A_I (4 panels) = 24 x 20 x 4 = 1920 sq ft

$L_r = 50(0.25 + 15/\sqrt{1920}) = 50(0.59) = 29.5$ psf

$w_u = 1.4(142) + 1.7(29.5) = 249$ psf

$V_u = 0.249(24 \times 20 - 1.97^2) = 118.6^k$

$\varphi V_c = 0.22\ A_c = 0.22(720.6) = 158.5^k > 118.6$ OK

per S.B.C. 1203.2
24x20 = 480 SF
R = 480x0.0008 = 38% < 40%
$R_{max} = 23.1\left(1 + \frac{D}{L}\right)$
= 89% > 38% ok
use 38%
L_r = 50 psf x 0.62
= 31 psf

Use 9 in. Slab

(3) Check limitations for slab analysis by coefficients

- 3 continuous spans in one direction, 5 in the other
- rectangular panels with long-to-short span ratio = 24/20 = 1.2 < 2
- successive span lengths in each direction are equal
- no offset columns
- LL/DL = 50/142 = 0.35 < 3
- slab system is without beams

(4) Factored moments in slab (NS-direction)

(a) Total panel moment M_o

clear span ≥ 0.65 o.c. span (ACI 13.6.2.5)

$$M_o = w_u \ell_2 \ell_n^2/8$$

$$= 0.278 \times 24 \times 18.83^2/8 = 295.7^{'k}$$

where w_u = 1.4(142) + 1.7(46.5*) = 278 psf

*live load reduction - A_I(one panel) = 24x20 = 480 sq ft

L_r = 50(0.25 + 15/√480) = 50(0.93) = 46.5 psf

S.B.C.

ℓ_2 = 24'-0

ℓ_n (interior span) = 20 - 1.33 = 18.67 ft

ℓ_n (end span) = 20 - 0.67 - 0.50 = 18.83 ft

Use larger value for both spans.

(b) Negative and positive factored moments

Division of the total panel moment M_o into negative and positive moments, and then, column and middle strip moments, involves direct application of the moment coefficients of Table 4-2. (p4-10)

Slab Moments ft-kips	End Spans			Interior Span
	Exterior Negative	Positive	Interior Negative	Positive
Total Moment	76.9	153.8	207.0	103.5
Column Strip	76.9	91.7	156.7	62.1
Middle Strip	0	62.1	50.3	41.4

Note: All negative moments are at face of column.

(5) Slab Reinforcement

Required slab reinforcement is easily determined using a tabular form as follows;

Span location		M_u	b[1]	d[2]	$A_s = \frac{M_u}{4d}$	$A_{s(min)}$[3]	No of #4 bars[4]	No of #5 bars[4]	Min.# Bars for crack control
END SPAN									
Column Strip	Ext. Negative	76.9	120	7.75	2.48	1.94	13	8	
	Positive	91.7	120	7.75	2.96	1.94	15	10	8
	Int. Negative	156.7	120	7.75	5.05	~~2.72~~ 1.94	26	17	
Middle Strip	Ext. Negative	0	168	7.75	--	2.72	14	11	
	Positive	62.1	168	7.75	2.00	2.72	14	11	11
	Int. Negative	50.3	168	7.75	1.62	2.72	14	11	
INTERIOR SPAN									
Column Strip	Positive	62.1	120	7.75	2.00	1.94	10	8	8
Middle Strip	Positive	41.4	168	7.75	1.34	2.72	14	11	11

Notes: [1]Column strip = $0.5\ell_2 \leq 0.5\ell_1$ = 0.5(20 x 12) = 120 in.

Middle strip = 24 x 12 - 120 = 168 in.

[2]Use average d = 9 - 1.25 = 7.75 in.

[3]$A_{s(min)}$ = 0.0018 bh = 0.0162 b

s_{max} = 2h < 18 = 2(9) = 18 in.

[4]Calculations

For s_{max}...120/18 = 6.7 spaces, say 8 bars

168/18 = 9.3 spaces, say 11 bars

For #4 bar...2.48/0.20 = 12.4 bars #5 bar....2.48/0.31 = 8 bars

2.96/0.20 = 14.8 bars 2.96/0.31 = 9.5 bars

5.05/0.20 = 25.3 bars 5.05/0.31 = 16.3 bars

2.72/0.20 = 13.6 bars 2.72/0.31 = 8.8 bars < 11

2.00/0.20 = 10 bars 2.00/0.31 = 6.5 bars < 8

(6) Check slab reinforcement at exterior column (12x12) for moment transfer between slab and column. For a slab without spandrel beams, the total exterior negative slab moment is resisted by the column strip.

$M_u = 76.9^{'k}$

Fraction transferred by flexure using ACI Eq. (13-1):

$M_u = 0.60(76.9) = 46.1^{'k}$ (p4-23)

$A_s = M_u/4d = 46.1/(4 \times 7.75) = 1.48 \text{ in.}^2$

For #4 bar = 1.48/0.20 = 7.4 bars <u>Say 7 #4 bars</u>

Must provide 7 #4 bars within an effective slab width (ACI 13.3.3.2) = $3h + c_2 = 3(9) + 12 = 39$ in.

Provide the required 7 #4 bars by concentrating 7 of the column strip bars (13 #4) within the 3'-3" slab width over the column. Distribute the other 6 column strip bars (3 each side) in the remaining column strip width. Check bar spacing:

For 7 #4 within 39 in. width ... 39/7 = 5.6 in.

For 6 #4 within (120 - 39) = 81 in. width ... 81/6 = 13.5 in. < 18 OK

No additional bars required for moment transfer.

(7) Selected bar layout and length details are shown below. Bar lengths are determined directly from Fig. 8-4. (p8-26)

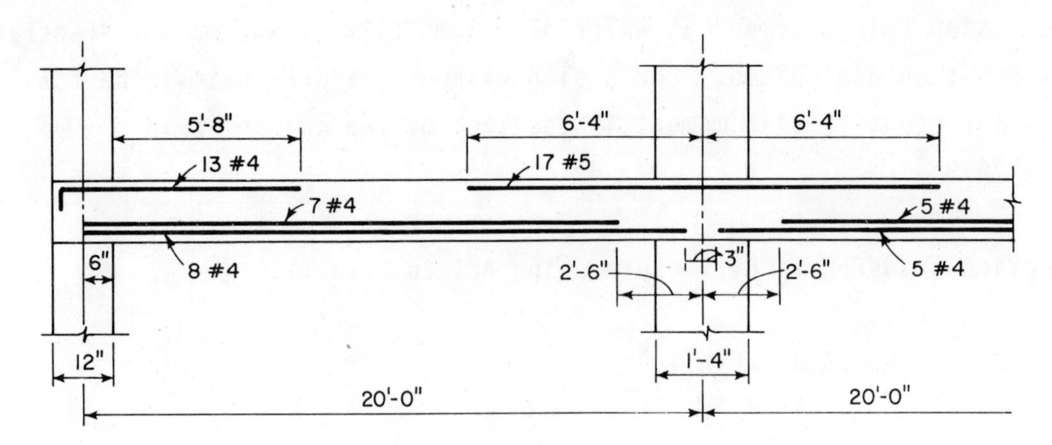

Column Strip

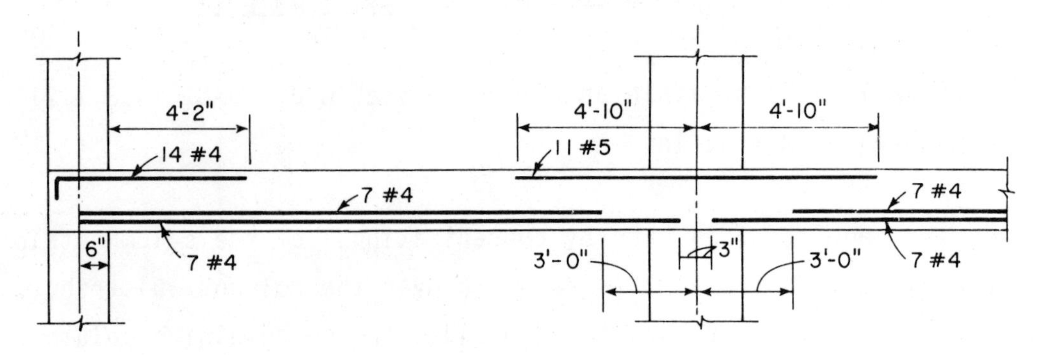

Middle Strip

Bar Length Details for Two-Way Flat Plate of Bldg. #2 - Alternate (2)
Interior Slab Panel (NS-Direction)

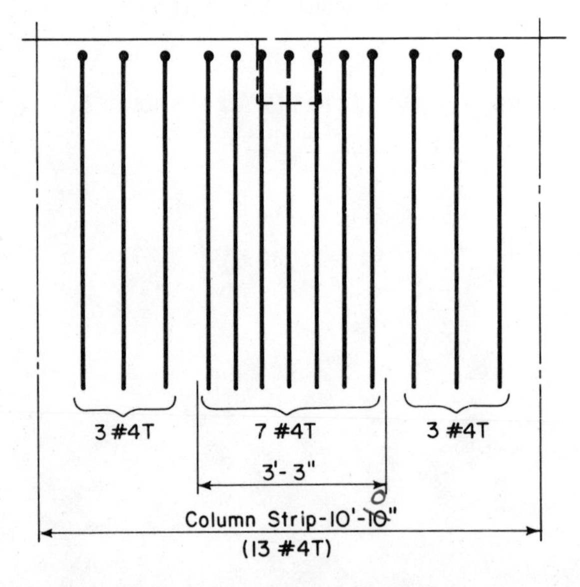

Bar Layout Detail for 13 #4 Top Bars at Ext. Columns

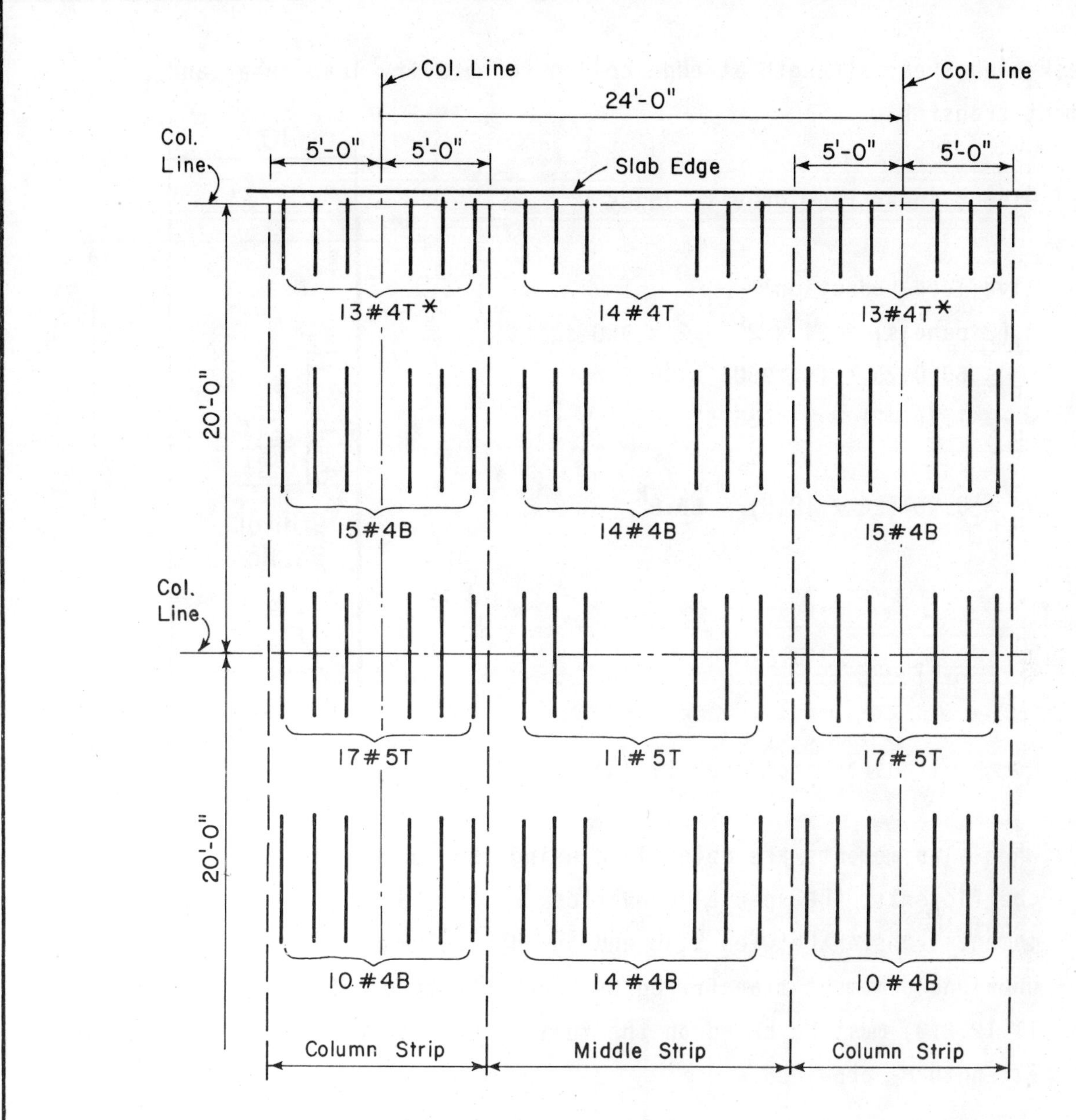

Bar Layout - Space Bars Uniformly within Each Column Strip and Middle Strip

(8) Check slab shear strength at edge column for gravity load shear and moment transfer.

(a) Direct shear from gravity loads

live load reduction:

A_I(2 panels) = 24 x 20 x 2 = 960 sq ft

$L_r = 50(0.25 + 15/\sqrt{960}) = 36.5$ psf

$w_u = 1.4(142) + 1.7(36.5) = 261$ psf

$V_u \simeq 0.261(24 \times 10.5) = 65.8^k$

SBC:
R = 24x10.5x0.0008
≈ 20% < 29%
Lr = 50x0.20 = 40 psf

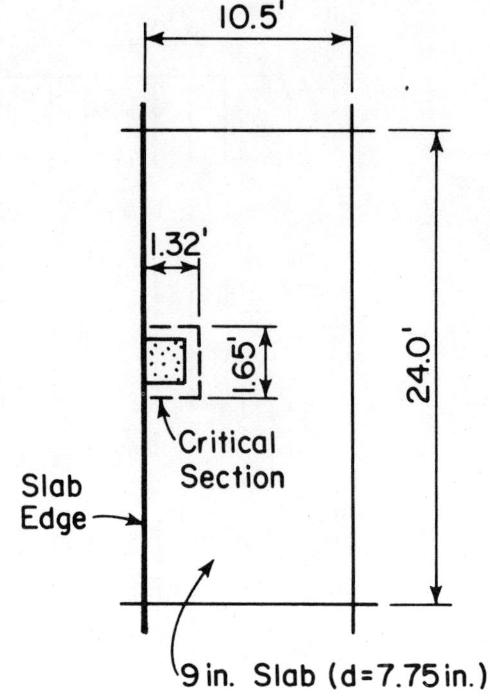

(b) Moment transfer from gravity loads

When slab moments are determined using the approximate moment coefficients, the special provisions of ACI 13.6.3.6 apply for moment transfer between slab and an edge column. The fraction of unbalanced moment transferred by eccentricity of shear (ACI 11.12.2.3) must be based on the full column strip nominal moment strength M_n provided. [40%]

For 13 #4 column strip bars, $A_s = 13(0.20) = 2.60$ in.^{2}

using the approximate expression for $A_s = M_u/4d$:

$$\varphi M_n = M_u = A_s(4d)$$

$$M_n = \frac{2.60(4 \times 7.75)}{0.9} = 89.6^{'k}$$

(c) Combined shear stress at inside face of critical transfer section

(pg 4-49)

From Table 4-9, for 9 in. slab with 12x12 column:

A_c = 390.8 in.^{2}

J/c = 2452 in.^{3}

direct — unbalanced moment (40% in shear)

$$v_u = V_u/A_c + \gamma_v M_u c/J$$

$$= 65{,}800/390.8 + 0.40 \times 89.6 \times 12000/2452$$

$$= 168.3 + 175.4 = 343.7 \text{ psi} \gg \varphi 4\sqrt{f'_c} = 215 \text{ psi}$$

The 9 in. slab is not adequate for shear and unbalanced moment transfer at the edge columns. Increase shear strength by providing drop panels at edge columns. Calculations not shown here.

4.7.2 - Example: Design a typical interior strip (NS-direction) of the two-way flat plate of Alternate (1) framing. Slab and column framing designed for both gravity and wind loads. Design interior strip for 1st-floor level (greatest wind load effects).

(1) Data: f'_c = 4000 psi

f_y = 60,000 psi

Floors: LL = 50 psf

DL = 136 psf

Preliminary sizing:

Slab = 8 1/2 in.

Columns int. = 16x16

ext. = 12x12

Spandrel beams = 12x20

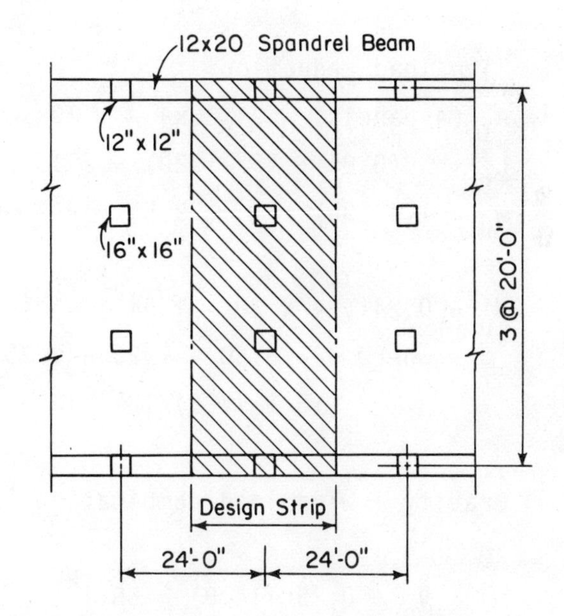

(2) Check slab thickness for deflection control and shear strength.

(a) Deflection control

From Table 4-1 (flat plate with spandrel beams):

$h = \ell_n/33 = (22.67 \times 12)/33 = 8.24$ in.

where $\ell_n = 24 - 16/12 = 22.67$ ft

(b) Shear strength. Check shear strength for 8 1/2 in. slab.

With the slab and column framing designed for both gravity and wind loads, slab shear strength needs to be checked for the combination of direct shear from gravity loads plus moment transfer from wind loads. Wind load analysis for Bldg. #2 is summarized in Fig. 2-14. Moment transfer between slab and column is greatest at the 1st-floor level where wind moment is largest. Transfer moment at 1st-floor level due to wind $M_w = 77.05 + 77.05 = 154.1^{'k}$

Direct shear from gravity loads:

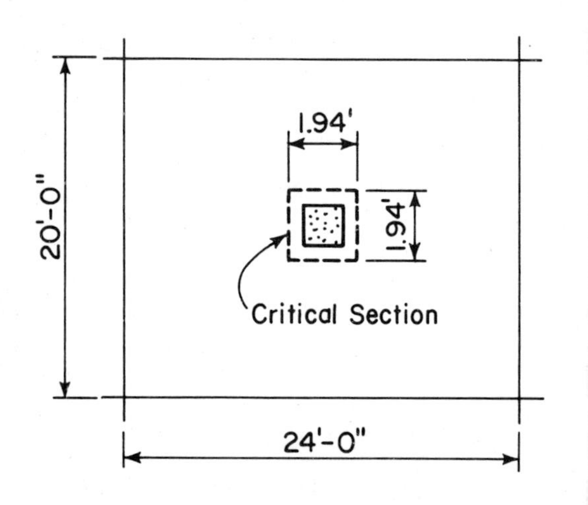

live load reduction:

A_I (4 panels) $= 24 \times 20 \times 4 = 1920$ sq ft

$L_r = 50(0.25 + 15/\sqrt{1920}) = 29.5$ psf

$w_u = 1.4(136) + 1.7(29.5) = 241$ psf

$V_u = 0.241(24 \times 20 - 1.94^2) = 114.8^k$

where $d = 8.50 - 1.25 = 7.25$ in.

$(16 + 7.25)/12 = 1.94$ ft

Gravity + wind load combination [ACI Eq. (9-2)]:

$V_u = 0.75(114.8) = 86.1^k$

$M_u = 0.75(1.7 \times 154.1) = 196.5^{'k}$

From Table 4-7, for 8-1/2 in. slab with 16x16 columns:

$A_c = 659.1$ in.^{2}

$J/c = 3201$ in.^{3}

Shear stress on critical transfer section:

$$v_u = V_u/A_c + \gamma_v M_u c/J$$
$$= 86{,}100/659.1 + 0.4 \times 196.5 \times 12000/3201$$
$$= 130.6 + 294.7 = 425.3 \text{ psi} \gg \varphi 4\sqrt{f'_c} = 215 \text{ psi}$$

The 8-1/2 in. slab is not adequate for gravity + wind load transfer at the interior columns.

Increase shear strength by providing drop panels at interior columns. Minimum slab thickness at drop panel (see Fig. 4-2) = 1.25(8.5) = 10.63 in. Dimension drop to actual lumber dimensions for economy of formwork. Try 2-1/4 in. drop.

$h = 8.5 + 2.25 = 10.75$ in. > 10.63.

$d = 7.25 + 2.25 = 9.5$ in.

Referring to Table 4-7:

$a = b = 16 + 9.5 = 25.5$ in. = 2.13 ft

$A_c = 4(25.5)9.5 = 969 \text{ in.}^2$

$J/c = [25.5 \times 9.5(25.5 \times 4) + 9.5^3]/3 = 8522 \text{ in.}^3$

$$v_u = 86{,}100/969 + 0.4 \times 196.5 \times 12{,}000/8522$$
$$= 88.9 + 110.7 = 199.6 \text{ psi} < 215 \quad \text{OK}$$

With drop panels, a lesser slab thickness for deflection control is permitted. From Table 4-1 (flat slab with spandrel beams): $h = \ell_n/36 = (22.67 \times 12)/36 = 7.56$ in. Could possibly reduce slab thickness from 8-1/2 to 8 in.; however, shear strength may not be adequate with the lesser slab thickness. For this example hold the slab thickness at 8-1/2 in. Note: the drop panels may not be required in the upper stories where the transfer moment due to wind becomes substantially less. See Fig. 2-14.

Use 8-1/2 in. slab
with 2-1/4 in. drop panels at interior columns of 1st story floor slab. Drop panel dimensions = $\ell/3$ = 24/3 = 8'-0". Use same dimension in both directions for economy of formwork.

(3) Factored moments in slab due to gravity load (NS-direction).

(a) Evaluate spandrel beam-to-slab stiffness ratio α and β_t.

Referring to Fig. 4-7:

ℓ = (20 x 12)/2 = 120

a = 20

b = 12

h = 8.5

a/h = 20/8.5 = 2.4

b/h = 12/8.5 = 1.4

f = 1.37

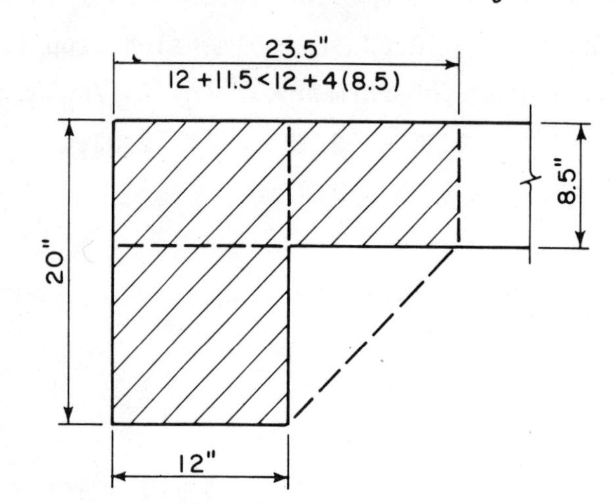

Spandrel Beam Section (See Fig. 4-5)

$$\alpha = \frac{b}{\ell}(a/h)^3 f = \frac{12}{120}(2.4)^3\ 1.37 = 1.89 > 0.8$$

Note Table 4-1: slab thickness $\ell_n/33$ OK for $\alpha > 0.8$.

$$\beta_t = \frac{C}{2I_s} = \frac{8525}{2(14740)} = 0.29 < 2.5$$

where $I_s = 24 \times 12(8.5)^3/12 = 14740$

C = larger value computed (with the aid of Fig. 4-8) for spandrel beam section.

x_1 = 8.5	x_2 = 11.5	x_1 = 12	x_2 = 8.5
y_1 = 23.5	y_2 = 12	y_1 = 20	y_2 = 11.5
C_1 = 3640*	C_2 = 2390	C_1 = 7165	C_2 = 1360
ΣC = 6030		ΣC = 8525	

*by interpolation (accurate enough)

(b) Total panel moment M_o

$$M_o = w_u \ell_2 \ell_n^2/8$$

$$= 0.269 \times 24 \times 18.83^2/8 = 286.1'^k$$

where w_u = 1.4(136) + 1.7(46.5*) = 269 psf

*live load reduction - A_I (one panel) = 24 x 20 = 480 sq ft

$L_r = 50(0.25 + 15/\sqrt{480}) = 50(0.93) = 46.5$ psf

ℓ_2 = 24'-0

ℓ_n (interior span) = 20 - 1.33 = 18.67 ft

ℓ_n (end span) = 20 - 0.67 - 0.50 = 18.83 ft

Use larger value for both spans.

(c) Negative and positive factored gravity load moments

Division of the total panel moment M_o into negative and positive moments, and then, column and middle strip moments involves direct application of the moment coefficients of Table 4-3. Note: the moment coefficients for the exterior negative column and middle strip moments need to be modified for β_t less than 2.5. For β_t = 0.29:

Column strip moment = $(0.30 - 0.03 \times 0.29)M_o = 0.29\ M_o$

Middle strip moment = $0.30\ M_o - 0.29\ M_o = 0.01\ M_o$

Slab moments ft-kips	End Spans			Interior Span
	Exterior negative	positive	Interior negative	positive
Total Moment	85.8	143.0	200.3	100.1
Column Strip	83.0	85.8	151.6	60.1
Middle Strip	2.8	57.2	48.7	40.0

Note: All negative moments are at face of column.

(4) Check negative moment sections for combined gravity + wind load moments.

(a) Exterior Negative

Consider wind load moments resisted by column strip as defined in Fig. 4-9.

gravity load only:

$M_u = 83.0^{'k}$ ACI Eq. (9-1)

gravity + wind loads:

$M_u = 0.75(83.0) + 0.75(1.7x77.05) = 160.5^{'k}$ (governs)

ACI Eq. (9-2)

also check for possible moment reversal due to wind moments.....

$M_u = 0.9(42) \pm 1.3(77.05) = -62.4^{'k}$ (reversal) ACI Eq. (9-3)

where $w_d = 136$ psf

$M_d = (0.136x24x18.83^2/8)0.29 = 42^{'k}$

(b) Interior Negative

gravity load only:

$M_u = 151.6^{'k}$

gravity + wind loads:

$M_u = 0.75(151.6) + 0.75(1.7x77.05) = 211.9^{'k}$ (governs)

and

$M_u = 0.9(76.8) \pm 1.3(77.05) = -31.1^{'k}$ (reversal)

where $M_d = 42(0.53/0.29) = 76.8^{'k}$

(5) Check slab section for moment strength

(a) At Exterior Negative

$b = 20\ M_u/d^2 = 20x160.5/7.25^2 = 61.1$ in. < 120 OK

where d = 8.5 - 1.25 = 7.25 in.

(b) At Interior Negative

$b = 20x211.9/9.50^2 = 47$ in. < 120 OK

where d = 7.25 + 2.25 = 9.50 in.

(6) Slab Reinforcement

Using a tabular form as follows:

Span location		M_u	b[1]	d[2]	$A_s = \frac{M_u}{4d}$	$A_{s(min)}$[3]	No of #4 bars[4]	No of #5 bars[4]
END SPAN								
Column Strip	Ext. Negative	160.5	120	7.25	5.53	1.84	28	18
		-62.4	120	7.25	2.15	-	11	9
	Positive	85.8	120	7.25	2.96	1.84	15	10
	Int. Negative	211.9	120	9.50	5.58	2.32	28	18
		-31.1	120	7.25	1.07	-	9	9
Middle Strip	Ext. Negative	2.8	168	7.25	0.10	2.57	13	11
	Positive	57.2	168	7.25	1.97	2.57	13	11
	Int. Negative	48.7	168	7.25	1.68	2.57	13	11
INTERIOR SPAN								
Column Strip	Positive	60.1	120	7.25	2.07	1.84	11	9
Middle Strip	Positive	40.0	168	7.25	1.38	2.57	9	9

Notes: [1]Column strip = $0.5\ell_2 \leq 0.5\ell_1$ = 0.5(20 x 12) = 120 in.

Middle strip = 24 x 12 - 120 = 168 in.

[2]Use average d = 8.5 - 1.25 = 7.25 in.

At drop panel d = 7.25 + 2.25 = 9.50 in. (negative only)

[3]$A_{s(min)}$ = 0.0018 bh

s_{max} = 2h < 18 = 2(8.5) = 17 in.

[4]Calculations

For s_{max}...120/17 = 7.1 spaces, say 9 bars

168/17 = 9.9 spaces, say 11 bars

For #4 bar...	#5 bar....
5.53/0.20 = 27.7 bars	5.53/0.31 = 17.8 bars
2.96/0.20 = 14.8 bars	2.96/0.31 = 9.5 bars
5.58/0.20 = 27.9 bars	5.58/0.31 = 18 bars
2.57/0.20 = 12.9 bars	2.57/0.31 = 8.3 bars < 11
2.07/0.20 = 10.4 bars	2.07/0.31 = 6.7 bars < 9
2.15/0.20 = 10.8 bars	2.15/0.31 = 6.9 bars < 9
1.07/0.20 = 5.4 bars < 9	1.07/0.31 = 3.5 bars < 9

(7) Check slab reinforcement at interior columns for moment transfer between slab and column. Shear strength of slab already checked for direct shear and moment transfer in Step (2)(b). Transfer moment at 1st-story due to wind $M_w = 154.1^{'k}$.

Fraction transferred by flexure using ACI Eqs. (13-1) and (9-3):

$$M_u = 0.60(1.3 \times 154.1) = 120.2^{'k}$$

$$A_s = M_u/4d = 120.2/4 \times 9.50 = 3.16 \text{ in.}^2$$

For #5 bar = 3.16/0.31 = 10.2 bars   Say 10 #5 bars

Must provide 10 #5 bars within an effective slab width = $3h + c_2$ = 3(10.75) + 16 = 48.3 in.

Provide the required 10 #5 bars by concentrating 10 of the column strip bars (18 #5) within the 4-ft slab width over the column. Distribute the other 8 column strip bars (4 each side) in the remaining column strip width. Check bar spacing:

48/9 spaces = ±5.3 in.

(120 - 48)/7 spaces = ±10.3 in. < 17   OK

Plan layout for the column strip bars at the interior slab-column supports, and the bar length details for the interior slab panel are shown below. Bar lengths for the middle strip are taken directly from Fig. 8-4. For the column strip, the bar lengths given in Fig. 8-4 (with drop panels) need to be modified to account for wind moment effects. In lieu of a rigorous analysis to determine bar-cutoffs based on a combination of gravity + wind moment variations, provide bar length details as follows:

For the top bars, cutoff 1/2 of the bars at $0.2\ell_n$ from supports and extend the remaining 1/2 full span length, with a Class C splice near center of span. Referring to Table 8-5, for #5 top bar and s > 6 in., Class C = 29 in. = 2'-5". At the exterior columns provide a 90° end-hook with 2 in.

minimum cover to edge of slab. From Table 8-2, for #5 bar, ℓ_{dh} = 9 in. < 12 - 2 = 10 in. OK. (For easier bar placement, alternate equal bar lengths at interior column supports.)

For the bottom bars, cutoff 6 #4 in the end spans at 0.125ℓ distance from the interior column, and extend 9 #4 a straight ℓ_d length beyond face of support. From Table 8-1, for #4 bar and s > 6 in., ℓ_d = 12 in. For the interior span extend the 11 #4 full span length with ℓ_d = 12 in. beyond face of support. At the exterior columns, provide a 90° end-hook with 2 in. minimum cover to edge of slab for all bottom bars.

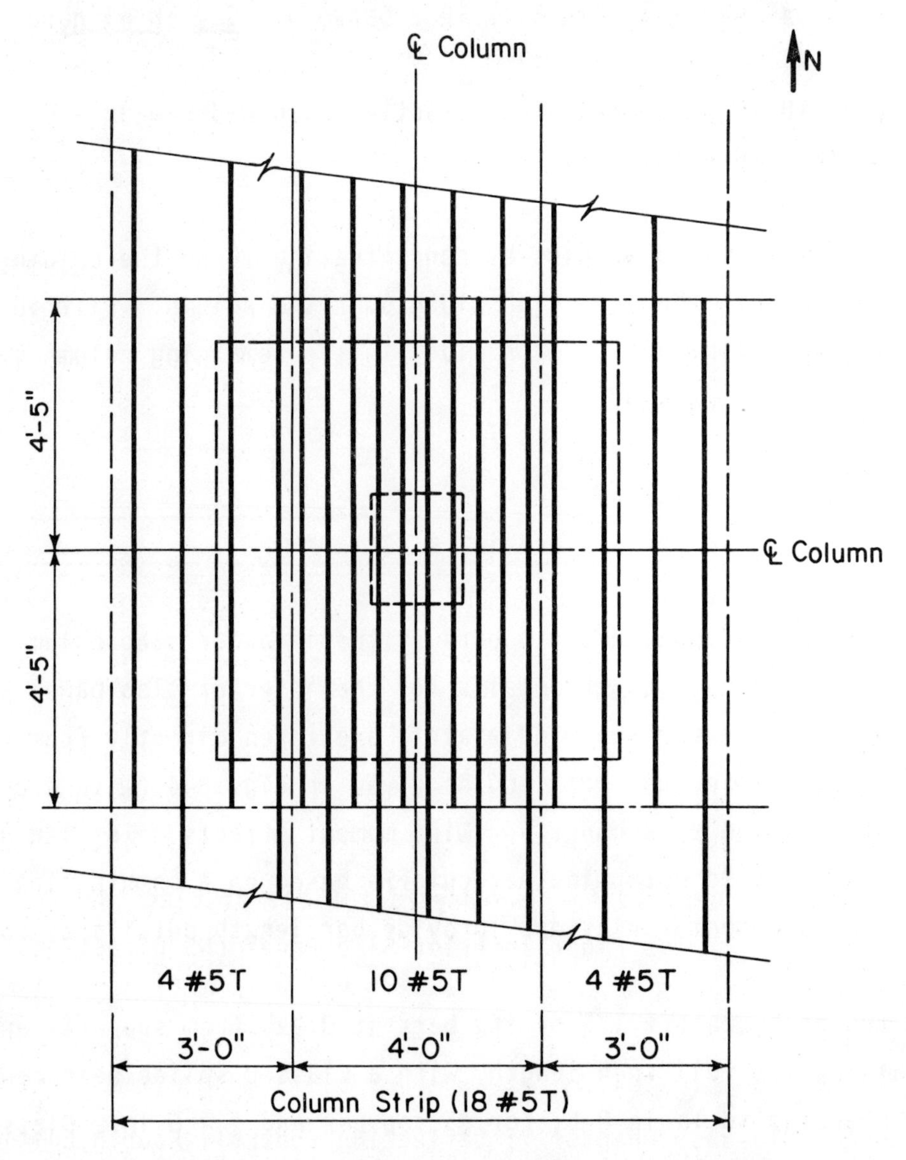

Bar Layout Detail for 18 #5 Top Bars at Interior Columns

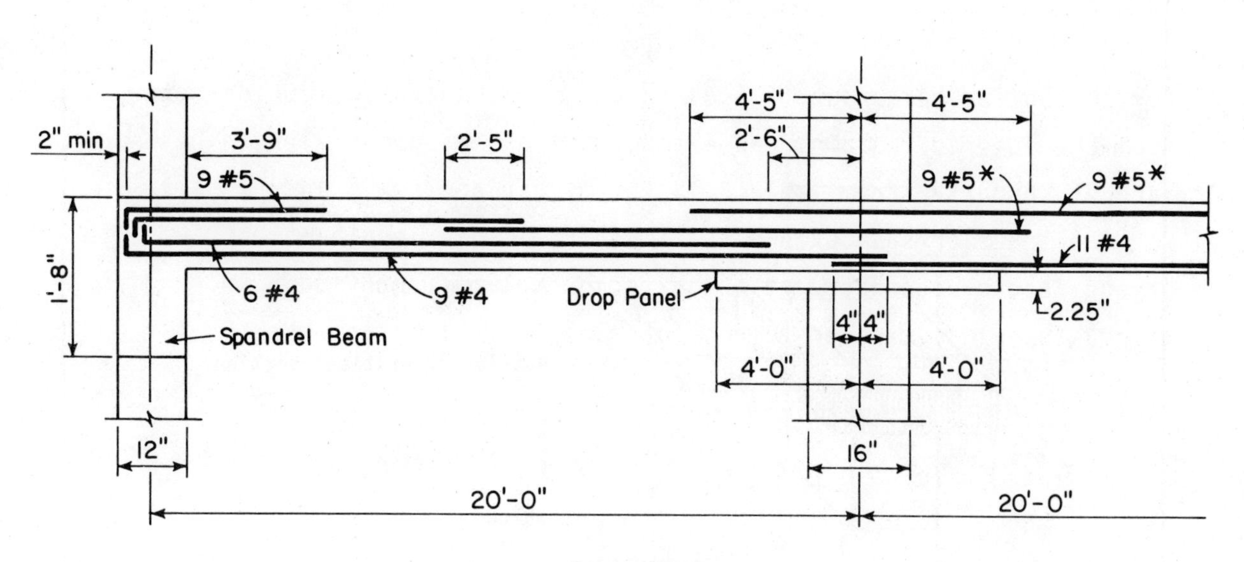

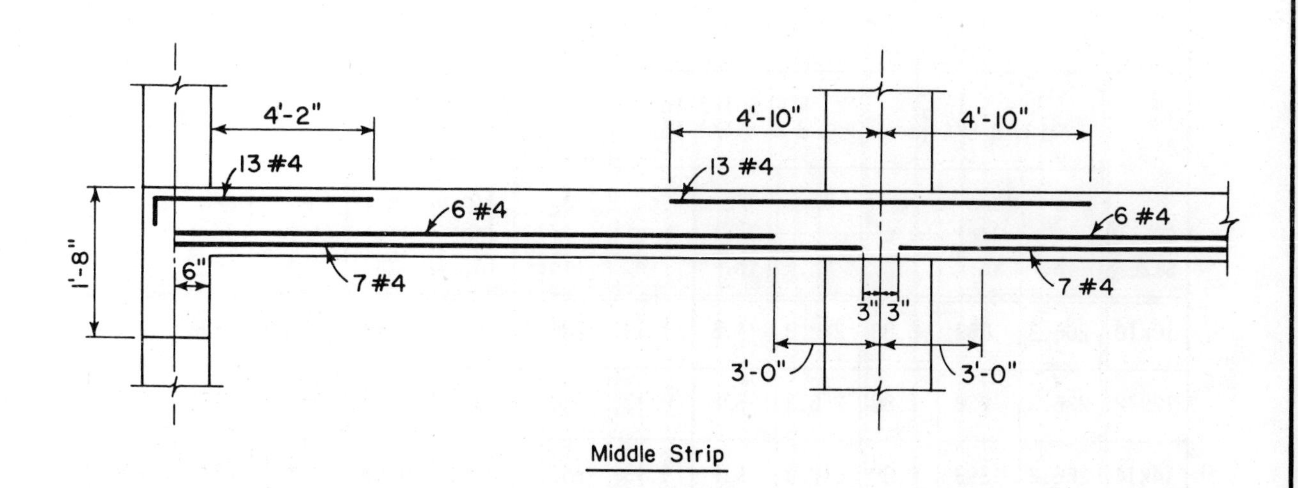

Bar Length Details for Two-Way Flat Plate of Bldg. #2 - Alternate (1) - 1st Floor Interior Slab Panel (NS-Direction)

Selected References

4.1 "Notes on ACI 318-83," Chapter 7: Deflection Control, Fourth Edition, EB070.04D, Portland Cement Association, Skokie, Ill.

Table 4-7 - Properties of Critical Transfer Section - INTERIOR COLUMN

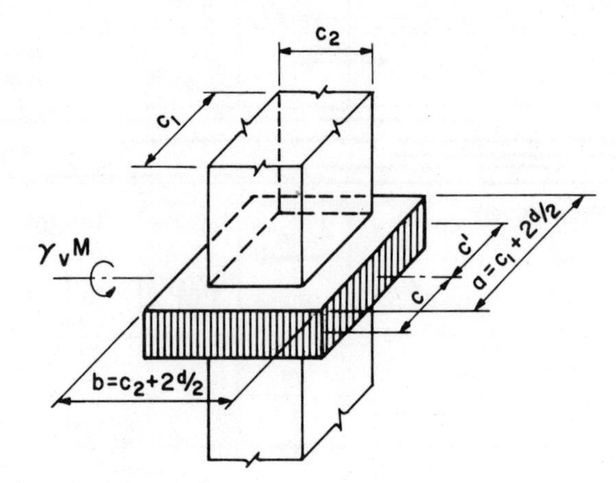

Concrete area of critical section:

$$A_c = 2(a + b)d$$

Modulus of critical section:

$$\frac{J}{c} = \frac{J}{c'} = [ad(a + 3b) + d^3]/3$$

where

$$c = c' = a/2$$

COL. SIZE	h = 5 in., d = 3 3/4 in.			h = 5 1/2 in., d = 4 1/4 in.			h = 6 in., d = 4 3/4 in.			h = 6 1/2 in., d = 5 1/4 in.		
	A_c in.^{2}	J/c = J/c' in.^{3}	c = c' in.	A_c in.^{2}	J/c = J/c' in.^{3}	c = c' in.	A_c in.^{2}	J/c = J/c' in.^{3}	c = c' in.	A_c in.^{2}	J/c = J/c' in.^{3}	c = c' in.
10x10	206.3	963	6.88	242.3	1176	7.13	280.3	1414	7.38	320.3	1676	7.63
12x12	236.3	1258	7.88	276.3	1522	8.13	318.3	1813	8.38	362.3	2131	8.63
14x14	266.3	1593	8.88	310.3	1913	9.13	356.3	2262	9.38	404.3	2642	9.63
16x16	296.3	1968	9.88	344.3	2349	10.13	394.3	2763	10.38	446.3	3209	10.63
18x18	326.3	2383	10.88	378.3	2831	11.13	432.3	3314	11.38	488.3	3832	11.63
20x20	356.3	2838	11.88	412.3	3358	12.13	470.3	3915	12.38	530.3	4511	12.63
22x22	386.3	3333	12.88	446.3	3930	13.13	508.3	4568	13.38	572.3	5246	13.63
24x24	416.3	3868	13.88	480.3	4548	14.13	546.3	5271	14.38	614.3	6037	14.63

Table 4-7 - Continued (INTERIOR COLUMN)

COL. SIZE	h = 7 in., d = 5 3/4 in.			h = 7 1/2 in., d = 6 1/8 in.			h = 8 in., d = 6 5/8 in.			h = 8 1/2 in., d = 7 1/8 in.		
	A_c $in.^2$	J/c = J/c' $in.^3$	c = c' in.	A_c $in.^2$	J/c = J/c' $in.^3$	c = c' in.	A_c $in.^2$	J/c = J/c' $in.^3$	c = c' in.	A_c $in.^2$	J/c = J/c' $in.^3$	c = c' in.
10x10	362.3	1965	7.88	395.1	2200	8.06	440.6	2538	8.31	488.1	2907	8.56
12x12	408.3	2479	8.88	444.1	2759	9.06	493.6	3161	9.31	545.1	3595	9.56
14x14	454.3	3054	9.88	493.1	3384	10.06	546.6	3855	10.31	602.1	4360	10.56
16x16	500.3	3690	10.88	542.1	4074	11.06	599.6	4619	11.31	659.1	3201	11.56
18x18	546.3	4388	11.88	591.1	4830	12.06	652.6	5453	12.31	716.1	6118	12.56
20x20	592.3	5147	12.88	640.1	5650	13.06	705.6	6359	13.31	773.1	7110	13.56
22x22	638.3	5967	13.88	689.1	6536	14.06	758.6	7335	14.31	830.1	8179	14.56
24x24	684.3	6849	14.88	738.1	7488	15.06	811.6	8382	15.31	887.1	9324	15.56

COL. SIZE	h = 9 in., d = 7 5/8 in.			h = 9 1/2 in., d = 8 in.			h = 10 in., d = 8 1/2 in.		
	A_c $in.^2$	J/c = J/c' $in.^3$	c = c' in.	A_c $in.^2$	J/c = J/c' $in.^3$	c = c' in.	A_c $in.^2$	J/c = J/c' $in.^3$	c = c' in.
10x10	537.6	3306	8.81	576	3627	9	629	4084	9.25
12x12	598.6	4063	9.81	640	4437	10	697	4968	10.25
14x14	659.6	4902	10.81	704	5333	11	765	5942	11.25
16x16	720.6	5822	11.81	768	6315	12	833	7008	12.25
18x18	781.6	6824	12.81	832	7381	13	901	8164	13.25
20x20	842.6	7906	13.81	896	8533	14	969	9410	14.25
22x22	903.6	9070	14.81	960	9771	15	1037	10748	15.25
24x24	964.6	10316	15.81	1024	11093	16	1105	12176	16.25

Table 4-8 - Properties of Critical Transfer Section - EDGE COLUMN - BENDING PARALLEL TO EDGE

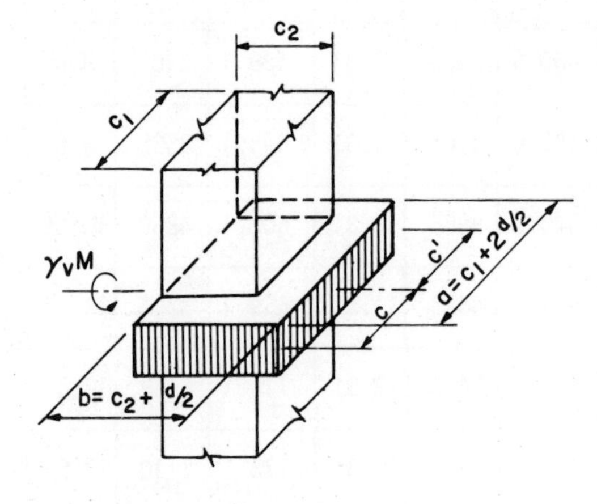

Concrete area of critical section:

$$A_c = (a + 2b)d$$

Modulus of critical section:

$$\frac{J}{c} = \frac{J}{c'} = [ad(a + 6b) + d^3]/6$$

where

$$c = c' = a/2$$

COL. SIZE	h = 5 in., d = 3 3/4 in.			h = 5 1/2 in., d = 4 1/4 in.			h = 6 in., d = 4 3/4 in.			h = 6 1/2 in., d = 5 1/4 in.		
	A_c $in.^2$	J/c = J/c' $in.^3$	c = c' in.	A_c $in.^2$	J/c = J/c' $in.^3$	c = c' in.	A_c $in.^2$	J/c = J/c' $in.^3$	c = c' in.	A_c $in.^2$	J/c = J/c' $in.^3$	c = c' in.
10x10	140.6	739	6.87	163.6	891	7.12	187.6	1057	7.37	212.6	1238	7.62
12x12	163.1	983	7.87	189.1	1175	8.12	216.1	1384	8.37	244.1	1609	8.62
14x14	185.6	1262	8.87	214.6	1499	9.12	244.6	1755	9.37	275.6	2029	9.62
16x16	208.1	1576	9.87	240.1	1863	10.12	273.1	2170	10.37	307.1	2497	10.62
18x18	230.6	1926	10.87	265.6	2267	11.12	301.6	2629	11.37	338.6	3015	11.62
20x20	253.1	2310	11.87	291.1	2710	12.12	330.1	3133	12.37	370.1	3581	12.62
22x22	275.6	2729	12.87	316.6	3192	13.12	358.6	3681	13.37	401.6	4197	13.62
24x24	298.1	3183	13.87	342.1	3715	14.12	387.1	4274	14.37	433.1	4861	14.62

Table 4-8 - Continued - (EDGE COLUMN - BENDING PARALLEL TO EDGE)

COL. SIZE	h = 7 in., d = 5 3/4 in.			h = 7 1/2 in., d = 6 1/8 in.			h = 8 in., d = 6 5/8 in.			h = 8 1/2 in. d = 7 1/8 in.		
	A_c in.^{2}	J/c = J/c' in.^{3}	c = c' in.	A_c in.^{2}	J/c = J/c' in.^{3}	c = c' in.	A_c in.^{2}	J/c = J/c' in.^{3}	c = c' in.	A_c in.^{2}	J/c = J/c' in.^{3}	c = c' in.
10x10	238.6	1435	7.87	258.8	1594	8.06	286.5	1820	8.31	315.3	2063	8.56
12x12	273.1	1852	8.87	295.5	2046	9.06	326.3	2321	9.31	358.0	2615	9.56
14x14	307.6	2322	9.87	332.3	2555	10.06	366.0	2884	10.31	400.8	3234	10.56
16x16	342.1	2846	10.87	369.0	3121	11.06	405.8	3508	11.31	443.5	3919	11.56
18x18	376.6	3423	11.87	405.8	3745	12.06	445.5	4195	12.31	486.3	4670	12.56
20x20	411.1	4054	12.87	442.5	4425	13.06	485.3	4943	13.31	529.0	5488	13.56
22x22	445.6	4739	13.87	479.3	5163	14.06	525.0	5733	14.31	571.8	6372	14.56
24x24	480.1	5477	14.87	516.0	5958	15.06	564.8	6626	15.31	614.5	7323	15.56

COL. SIZE	h = 9 in., d = 7 5/8 in.			h = 9 1/2 in., d = 8 in.			h = 10 in., d = 8 1/2 in.		
	A_c in.^{2}	J/c = J/c' in.^{3}	c = c' in.	A_c in.^{2}	J/c = J/c' in.^{3}	c = c' in.	A_c in.^{2}	J/c = J/c' in.^{3}	c = c' in.
10x10	345.0	2325	8.81	368.0	2533	9.00	399.5	2828	9.25
12x12	390.8	2930	9.81	416.0	3179	10.00	450.5	3529	10.25
14x14	436.5	3605	10.81	464.0	3899	11.00	501.5	4310	11.25
16x16	482.3	4352	11.81	512.0	4693	12.00	552.5	5170	12.25
18x18	528.0	5170	12.81	560.0	5563	13.00	603.5	6109	13.25
20x20	573.8	6060	13.81	608.0	6507	14.00	654.5	7128	14.25
22x22	619.5	7020	14.81	656.0	7525	15.00	705.5	8226	15.25
24x24	665.3	8052	15.81	704.0	8619	16.00	756.5	9403	16.25

Table 4-9 - Properties of Critical Transfer Section - EDGE COLUMN - BENDING PERPENDICULAR TO EDGE

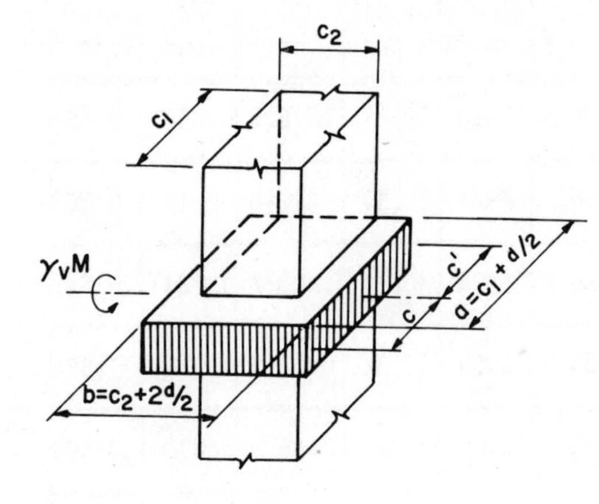

Concrete area of critical section:

$A_c = (2a + b)d$

Modulus of critical section:

$\frac{J}{c} = [2ad(a + 2b) + d^3(2a + b)/a]/6$

$\frac{J}{c'} = [2a^2d(a + 2b) + d^3(2a + b)]/6(a + b)$

where

$c = a^2/2(a + b)$

$c' = a(a + b)/(2a + b)$

	h = 5 in., d = 3 3/4 in.					h = 5 1/2 in., d = 4 1/4 in.					h = 6 in., d = 4 3/4 in.				
COL. SIZE	A_c in.^{2}	J/c in.^{3}	J/c' in.^{3}	c in.	c' in.	A_c in.^{2}	J/c in.^{3}	J/c' in.^{3}	c in.	c' in.	A_c in.^{2}	J/c in.^{3}	J/c' in.^{3}	c in.	c' in.
10x10	140.6	612	284	3.76	8.11	163.6	738	339	3.81	8.30	187.6	878	400	3.87	8.49
12x12	163.1	815	381	4.42	9.44	189.1	973	453	4.48	9.64	216.1	1146	529	4.54	9.83
14x14	185.6	1047	494	5.09	10.78	214.6	1242	583	5.14	10.97	244.6	1453	677	5.20	11.16
16x16	208.1	1309	622	5.75	12.11	240.1	1545	730	5.81	12.31	273.1	1798	844	5.87	12.50
18x18	230.6	1602	765	6.42	13.45	265.6	1882	894	6.48	13.64	301.6	2181	1030	6.53	13.83
20x20	253.1	1924	923	7.08	14.78	291.1	2253	1075	7.14	14.97	330.1	2602	1235	7.20	15.17
22x22	275.6	2277	1095	7.75	16.11	316.6	2658	1273	7.81	16.31	358.6	3061	1459	7.86	16.50
24x24	298.1	2659	1283	8.42	17.45	342.1	3097	1488	8.47	17.64	387.1	3558	1702	8.53	17.83

Table 4-9 - Continued - (EDGE COLUMN - BENDING PERPENDICULAR TO EDGE)

	h = 6 1/2 in., d = 5 1/4 in.					h = 7 in., d = 5 3/4 in.					h = 7 1/2 in., d = 6 1/8 in.				
COL. SIZE	A_c in.^{2}	J/c in.^{3}	J/c' in.^{3}	c in.	c' in.	A_c in.^{2}	J/c in.^{3}	J/c' in.^{3}	c in.	c' in.	A_c in.^{2}	J/c in.^{3}	J/c' in.^{3}	c in.	c' in.
10x10	212.6	1030	467	3.93	8.68	238.6	1197	538	3.99	8.88	258.8	1332	596	4.03	9.02
12x12	244.1	1334	612	4.59	10.02	273.1	1537	701	4.65	10.21	295.5	1701	772	4.70	10.36
14x14	275.6	1680	779	5.26	11.36	307.6	1924	886	5.32	11.55	332.3	2118	972	5.36	11.69
16x16	307.1	2068	966	5.92	12.69	342.1	2356	1095	5.98	12.88	369.0	2585	1196	6.03	13.03
18x18	338.6	2498	1174	6.59	14.02	376.6	2835	1326	6.65	14.22	405.8	3101	1445	6.69	14.36
20x20	370.1	2970	1404	7.26	15.36	411.1	3360	1581	7.31	15.55	442.5	3666	1719	7.36	15.70
22x22	401.6	3485	1654	7.92	16.69	445.6	3931	1858	7.98	16.89	479.3	4280	2017	8.02	17.03
24x24	433.1	4041	1926	8.59	18.03	480.1	4548	2158	8.64	18.22	516.0	4943	2339	8.69	18.36

	h = 8 in., d = 6 5/8 in.					h = 8 1/2 in., d = 7 1/8 in.					h = 9 in., d = 7 5/8 in.				
COL. SIZE	A_c in.^{2}	J/c in.^{3}	J/c' in.^{3}	c in.	c' in.	A_c in.^{2}	J/c in.^{3}	J/c' in.^{3}	c in.	c' in.	A_c in.^{2}	J/c in.^{3}	J/c' in.^{3}	c in.	c' in.
10x10	286.5	1526	679	4.09	9.21	315.3	1737	768	4.15	9.40	345.0	1964	863	4.21	9.59
12x12	326.3	1933	872	4.76	10.55	358.0	2184	980	4.81	10.74	390.8	2452	1094	4.87	10.93
14x14	366.0	2394	1092	5.42	11.88	400.8	2688	1220	5.48	12.07	436.5	3002	1356	5.54	12.27
16x16	405.8	2907	1339	6.08	13.22	443.5	3250	1489	6.14	13.41	482.3	3613	1648	6.20	13.60
18x18	445.5	3474	1612	6.75	14.55	486.3	3868	1787	6.81	14.75	528.0	4285	1970	6.87	14.94
20x20	485.3	4094	1911	7.41	15.89	529.0	4544	2113	7.47	16.08	573.8	5019	2323	7.53	16.27
22x22	525.0	4767	2237	8.08	17.22	571.8	5278	2467	8.14	17.41	619.5	5813	2707	8.20	17.61
24x24	564.8	5493	2589	8.75	18.56	614.5	6068	2850	8.80	18.75	665.3	6669	3121	8.86	18.94

Table 4-9 - Continued - (EDGE COLUMN - BENDING PERPENDICULAR TO EDGE)

	h = 9 1/2 in., d = 8 in.					h = 10 in., d = 8 1/2 in.				
COL. SIZE	A_c in.^{2}	J/c in.^{3}	J/c' in.^{3}	c in.	c' in.	A_c in.^{2}	J/c in.^{3}	J/c' in.^{3}	c in.	c' in.
10x10	368.0	2147	939	4.26	9.73	399.5	2407	1047	4.32	9.92
12x12	416.0	2667	1185	4.92	11.07	450.5	2970	1313	4.98	11.26
14x14	464.0	3251	1463	5.58	12.41	501.5	3601	1613	5.64	12.60
16x16	512.0	3900	1773	6.25	13.75	552.5	4302	1947	6.30	13.94
18x18	560.0	4613	2114	6.91	15.08	603.5	5071	2314	6.97	15.27
20x20	608.0	5390	2488	7.57	16.42	654.5	5908	2716	7.63	16.61
22x22	656.0	6232	2893	8.24	17.75	705.5	6813	3151	8.30	17.94
24x24	704.0	7138	3331	8.90	19.09	756.5	7786	3621	8.96	19.28

Table 4-10 - Properties of Critical Transfer Section - CORNER COLUMN

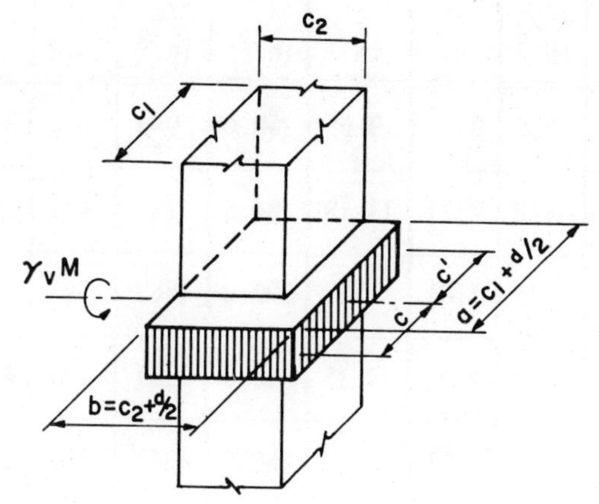

Concrete area of critical section:

$A_c = (a + b)d$

Modulus of critical section:

$\frac{J}{c} = [ad(a + 4b) + d^3(a + b)/a]/6$

$\frac{J}{c'} = [a^2d(a + 4b) + d^3(a + b)]/6(a + 2b)$

where

$c = a^2/2(a + b)$

$c' = a(a + 2b)/2(a + b)$

COL. SIZE	h = 5 in., d = 3 3/4 in.					h = 5 1/2 in., d = 4 1/4 in.					h = 6 in., d = 4 3/4 in.				
	A_{c_2} in.^{2}	J/c in.^{3}	J/c' in.^{3}	c in.	c' in.	A_{c_2} in.^{2}	J/c in.^{3}	J/c' in.^{3}	c in.	c' in.	A_{c_2} in.^{2}	J/c in.^{3}	J/c' in.^{3}	c in.	c' in.
10x10	89.1	458	153	2.96	8.90	103.1	546	182	3.03	9.09	117.6	642	214	3.09	9.28
12x12	104.1	619	206	3.46	10.40	120.1	732	244	3.53	10.59	136.6	854	285	3.59	10.78
14x14	119.1	805	268	3.96	11.90	137.1	946	315	4.03	12.09	155.6	1097	366	4.09	12.28
16x16	134.1	1016	339	4.46	13.40	154.1	1189	396	4.53	13.59	174.6	1372	457	4.59	13.78
18x18	149.1	1252	417	4.96	14.90	171.1	1460	487	5.03	15.09	193.6	1679	560	5.09	15.28
20x20	164.1	1513	504	5.46	16.40	188.1	1759	586	5.53	16.59	212.6	2017	672	5.59	16.78
22x22	179.1	1799	600	5.96	17.90	205.1	2087	696	6.03	18.09	231.6	2388	796	6.09	18.28
24x24	194.1	2110	703	6.46	19.40	222.1	2443	814	6.53	19.59	250.6	2789	930	6.59	19.78

Table 4-10 - Continued - (CORNER COLUMN)

COL. SIZE	h = 6 1/2 in., d = 5 1/4 in.					h = 7 in., d = 5 3/4 in.					h = 7 1/2 in., d = 6 1/8 in.				
	A_{c2} in.^{2}	J/c in.^{3}	J/c' in.^{3}	c in.	c' in.	A_{c2} in.^{2}	J/c in.^{3}	J/c' in.^{3}	c in.	c' in.	A_{c2} in.^{2}	J/c in.^{3}	J/c' in.^{3}	c in.	c' in.
10x10	132.6	746	249	3.15	9.46	148.1	858	286	3.21	9.65	160.0	948	316	3.26	9.79
12x12	153.6	984	328	3.65	10.96	171.1	1124	375	3.71	11.15	184.5	1235	412	3.76	11.29
14x14	174.6	1257	419	4.15	12.46	194.1	1428	476	4.21	12.65	209.0	1563	521	4.26	12.79
16x16	195.6	1566	522	4.65	13.96	217.1	1770	590	4.71	14.15	233.5	1931	644	4.76	14.29
18x18	216.6	1909	636	5.15	15.46	240.1	2151	717	5.21	15.65	258.0	2341	780	5.26	15.79
20x20	237.6	2288	763	5.65	16.96	263.1	2571	857	5.71	17.15	282.5	2791	930	5.76	17.29
22x22	258.6	2701	900	6.15	18.46	286.1	3028	1009	6.21	18.65	307.0	3283	1094	6.26	18.79
24x24	279.6	3150	1050	6.65	19.96	309.1	3524	1175	6.71	20.15	331.5	3815	1272	6.76	20.29

COL. SIZE	h = 8 in., d = 6 5/8 in.					h = 8 1/2 in., d = 7 1/8 in.					h = 9 in., d = 7 5/8 in.				
	A_{c2} in.^{2}	J/c in.^{3}	J/c' in.^{3}	c in.	c' in.	A_{c2} in.^{2}	J/c in.^{3}	J/c' in.^{3}	c in.	c' in.	A_{c2} in.^{2}	J/c in.^{3}	J/c' in.^{3}	c in.	c' in.
10x10	176.4	1075	358	3.32	9.98	193.3	1213	404	3.39	10.17	210.6	1360	453	3.45	10.35
12x12	202.9	1391	464	3.82	11.48	221.8	1559	520	3.89	11.67	241.1	1737	579	3.95	11.85
14x14	229.4	1752	584	4.32	12.98	250.3	1952	651	4.39	13.17	271.6	2164	721	4.45	13.35
16x16	255.9	2156	719	4.82	14.48	278.8	2393	798	4.89	14.67	302.1	2642	881	4.95	14.85
18x18	282.4	2605	868	5.32	15.98	307.3	2881	960	5.39	16.17	332.6	3171	1057	5.45	16.35
20x20	308.9	3097	1032	5.82	17.48	335.8	3417	1139	5.89	17.67	363.1	3751	1251	5.95	17.85
22x22	335.4	3634	1211	6.32	18.98	364.3	4000	1333	6.39	19.17	393.6	4381	1460	6.45	19.35
24x24	361.9	4215	1405	6.82	20.48	392.8	4631	1544	6.89	20.67	424.1	5063	1688	6.95	20.85

Table 4-10 - Continued - (CORNER COLUMN)

	h = 9 1/2 in., d = 8 in.					h = 10 in., d = 8 1/2 in.				
COL. SIZE	A_c in.^{2}	J/c in.^{3}	J/c' in.^{3}	c in.	c' in.	A_c in.^{2}	J/c in.^{3}	J/c' in.^{3}	c in.	c' in.
10x10	224.0	1477	492	3.50	10.50	242.3	1643	548	3.56	10.68
12x12	256.0	1877	626	4.00	12.00	276.3	2075	692	4.06	12.18
14x14	288.0	2331	777	4.50	13.50	310.3	2564	855	4.56	13.68
16x16	320.0	2837	946	5.00	15.00	344.3	3109	1036	5.06	15.18
18x18	352.0	3397	1132	5.50	16.50	378.3	3711	1237	5.56	16.68
20x20	384.0	4011	1337	6.00	18.00	412.3	4370	1457	6.06	18.18
22x22	416.0	4677	1559	6.50	19.50	446.3	5086	1695	6.56	19.68
24x24	448.0	5397	1799	7.00	20.00	480.3	5858	1953	7.06	21.18

5

Simplified Design for Columns

Larry W. Cole*

5.1 - INTRODUCTION

Use of strength design methods combined with higher strength materials has had a significant effect on the design and use of concrete columns. Larger bay sizes with fewer more heavily loaded columns has resulted. Increased use of higher strength concretes has also resulted in columns of smaller size and, therefore, higher column slenderness. These factors have combined to make column design more complex and/or complicated in today's concrete structures.

Yet, for buildings with adequate shear walls, column design can be considerably simplified. In some structures, especially low-rise buildings, it may not be desirable or economical to include structural wall bracing and the frame must stand alone to resist both gravity and lateral load effects as an unbraced frame. Although the design of columns in frames considered unbraced cannot be as easily simplified as for the braced condition, some simplification is possible, especially for building frames of moderate size and height with regular framing layouts. Design simplifications for both the "braced" and "unbraced" columns are presented in this chapter, with emphasis on a simplified approach for proportioning the more complex and complicated of the two, the unbraced slender column.

*Regional Structural Engineer, Rocky Mountain-Northwest Region, PCA

5.2 - DESIGN CONSIDERATIONS

5.2.1 - Column Size

Floor bay sizes determine the total tributary load to a column. Larger bay sizes mean more load to each column. Bay size is often dictated by the architectural and functional requirements of the building. Large bay sizes may be required to achieve maximum unobstructed floor space. The floor system used may also dictate the column spacing; for example, the economical use of a flat plate floor system requires a lesser column spacing than a pan joist floor system. Architecturally, larger column sizes can give the impression of solidity and strength, whereas smaller columns can express slender grace. Aside from architectural considerations, there are three goals a designer would like to achieve: adequate strength, an economical column, and design simplicity. Minimum column size and concrete cover to reinforcement may be governed by fire-resistance criteria; see Chapter 10, Tables 10-2 and 10-5.

5.2.2 - Column Constructability

Columns must be sized not only for adequate strength, but also for constructability. For proper concrete placement and consolidation, the designer must select column size and reinforcement to ensure that the column reinforcement is not congested. Bar lap splices and location of bars in connecting beams and slabs must be considered. Columns designed with a smaller number of larger bars improve constructability.

5.2.3 - Column Economics

Concrete is more cost-effective than reinforcement for carrying axial load; thus, economics of concrete columns (material and labor) generally dictate larger column sizes with lesser amounts of reinforcement. Also, columns designed with a smaller number of larger bars are more economical than columns designed with a larger number of smaller bars.

Reuse of column forms from story level to story level allows for more economical construction. It is economically unsound to vary column size to suit the load on each story level. Greater economies are achieved, especially for buildings of moderate height, by maintaining the same column size for the entire building height, and varying the reinforcement for each story level.

5.3 - SIMPLIFIED DESIGN FOR COLUMNS

Numerous design aids and computer programs are available for determining size and reinforcement of columns employed in building construction. Tables, charts, and graphs provide design data for a wide variety of column sizes and shapes, reinforcement layouts, load eccentricities and other variables, to eliminate the necessity for making complex and repetitious calculations to determine the capacities of trial column sizes. Three such design aids are presented in References 5.1, 5.2, and 5.3. In addition, extensive column load tables are available in the CRSI Handbook.[5.4] Each publication presents the design data in a somewhat different format, but study of their accompanying explanatory material will readily reveal their method of use.

For the purpose of this manual, addressing buildings of moderate size and height, design charts for a range of typical square columns with four symmetrical bar arrangements are presented in Figs. 5-6 through 5-14 at the back of this chapter. To allow rapid selection of column size and reinforcement required for a given loading (P_u and M_u), the design charts are plotted for specific column sizes and numbers of reinforcing bars. All charts are based on the 4000/60,000 material strength combination, with the material strength reduction ($\varphi = 0.70$) for tied columns included in the design charts so that design strengths are obtained directly ($P_u = \varphi P_n$ and $M_u = \varphi M_n$). All bar arrangements are symmetrical as shown in Fig. 5-1, with 1-1/2 in. cover to #3 ties for #10 bar and smaller and 2 in. cover to #4 tie for #11 bars. **#14 and #18 bars are not considered due to limited availability.**

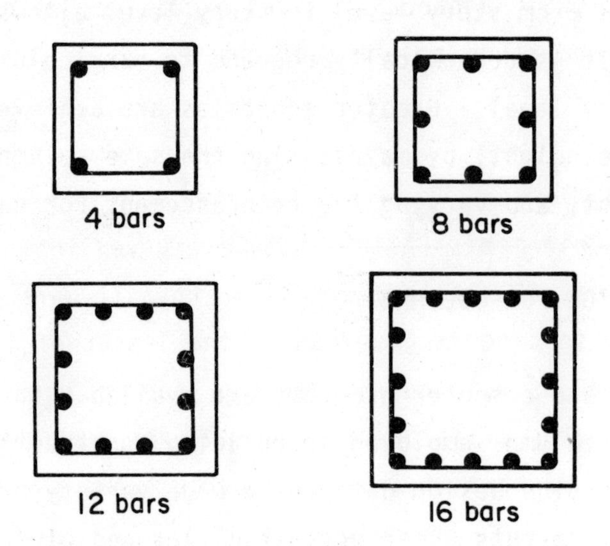

Fig. 5-1 Bar Arrangements for Column Design Charts

For each capacity curve, the number and bar size, and reinforcement percentage ($\rho_t = 100\ A_{st}/A_g$) are noted. For simplification, the design curves are plotted as straight lines between the critical points. For each design curve, the maximum axial load capacity is limited by ACI Eq. (10-2)...$P_u \leq \varphi\ (0.80P_o)$. Columns are often intentionally made larger than necessary for formwork economy, using the same size column from footing to roof of a multicolumn stack. When a column has a larger cross section than required for load considerations, a reduced effective area not less than one-half the total area may be used for determining reinforcement area and load capacity (ACI 10.8.4). This code provision permits an economical minimum steel percentage of 0.5% for these types of columns, instead of the usual minimum of 1% required by ACI 10.9.1. Accordingly, each column chart includes design curves for reinforcing bar arrangements with percentages less than 1%. The capacity curves for $\rho < 1\%$ are based on a reduced effective area to provide 1% reinforcement for the respective bar areas.[5.5]

The designer can easily develop additional design charts for other column sizes and material strengths. Should rectangular or round columns be preferred, design graphs presented in Reference 5.2 may be used; the graphs are nondimensional and cover an extensive range of column shapes and material strengths. Also, the CRSI Handbook[5.4] gives extensive design data for square, rectangular, and round column sections.

5.3.1 Column Ties

Column tie spacing requirements (ACI 7.10.5) are given in Table 5-1. For #10 column bars and smaller, #3 ties are required; for #11 bars, a #4 tie must be used. Maximum tie spacing is governed by the lesser of three criteria; 16 d_b (column bars), 48 d_b (tie bars), and the least column dimension. The lesser of the bar diameter criteria is tabulated for required minimum tie sizes and sizes of column bars; however, in no case can these ties be spaced further apart than the column dimension.

Table 5-1 Column Tie Spacing

Tie Size	Column Bars	Max. Spacing*
#3	# 4	8
	# 5	10
	# 6	12
	# 7	14
	# 8	16
	# 9	18
	#10	18
#4	#11	22

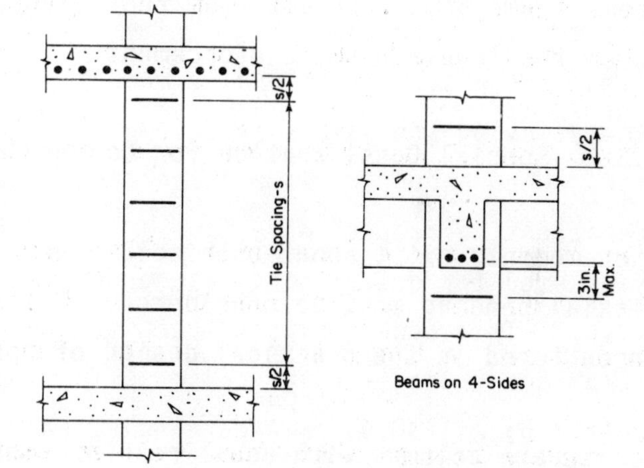

*Max. spacing not to exceed least column dimension.

Suggested tie details at each tie spacing to satisfy ACI 7.10.5.3 are shown in Fig. 5-2 for the 8, 12, and 16 column bar arrangements included in Figs. 5-6 through 5-14. In any square (or rectangular) bar arrangement, the four corner bars are preferably enclosed by a single one-piece tie to aid in holding the bars in position during construction. The one-piece tie is formed by overlapping 90° standard tie end hooks. In placing successive sets of ties, position of end hooks should be alternated by rotating the tie 90° or 180°. A note to this effect should be added to the detail on the structural drawings and should be repeated by the detailers on the placing drawings. For the 8 bar and 16 bar arrangement, the intermediate bars required to have support by a tie corner can most easily be supported by a "candy-stick" tie with 90° and 180° opposite end hooks (for easy field erection). Again, the designer should add the note to alternate the position of the 90° and 180° hooks at each successive tie spacing. The two-piece tie shown for the 12 bar arrangement should be lap spliced 12 in. minimum. To eliminate the supplementary ties for the 8, 12, and 16 bar arrangements, 2, 3, and 4 bar bundles of each corner may also be used.

Column ties must be located not more than one-half a tie spacing above top of footing or slab in any story, and not more than one-half a tie spacing below the lowest reinforcement in the slab (or drop panel) above. Where beams frame into a column from four sides, ties may be terminated 3 in. below the lowest beam reinforcement.

5.3.2 Special Consideration for Corner Columns

When moments exist about both column axis simultaneously with axial load, biaxial bending must be considered. This special problem is often encountered in the practical design of corner columns.

For square columns with equal reinforcement in all four faces, the uniaxial moment capacity for any biaxial loading condition can be approximated (conservatively) by the vector summation of the uniaxial capacities; see Fig. 5-3. For a rapid practical solution, simply add the uniaxial design moments:

$$M_u = M_{ux} + M_{uy}$$

The square column is then designed as one subject to uniaxial loading (P_u and M_u) using design charts of Figs. 5-6 through 5-14.

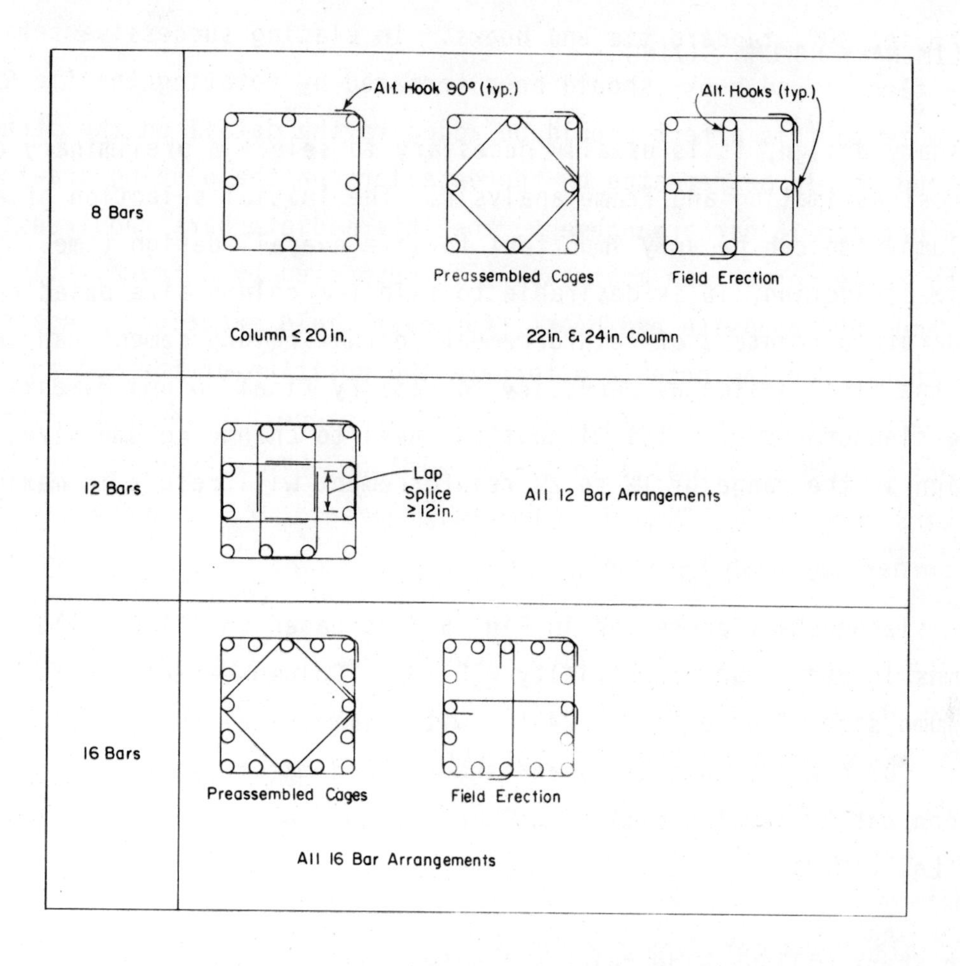

Fig. 5-2 Column Tie Details

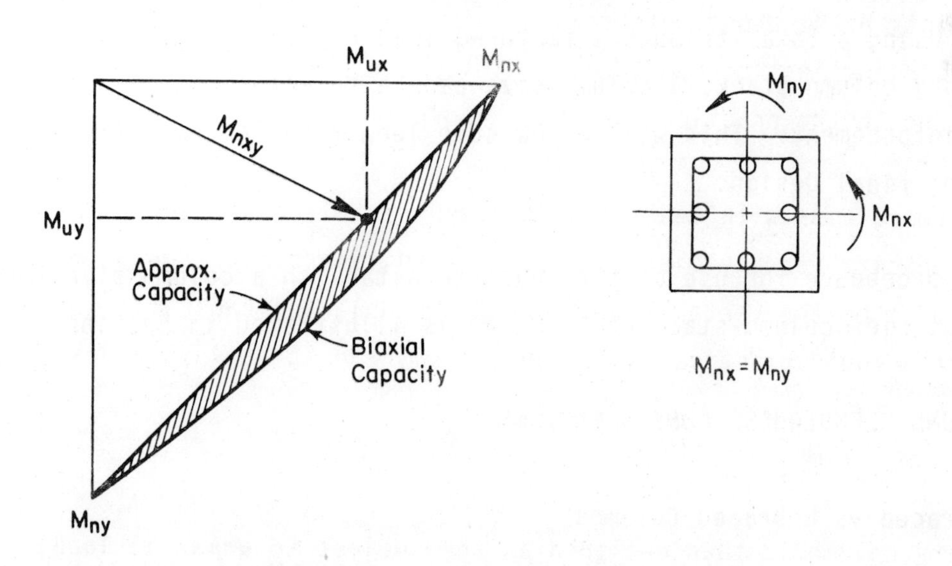

Fig. 5-3 Biaxial Capacity Approximation for Square Column with Symmetrical Reinforcement

5.4 - PRELIMINARY COLUMN SIZING

In preliminary design, it is usually necessary to select a preliminary column size for cost estimating and frame analysis. The initial selection of a usable column size can be very important to save overall design time. For initial size selection, it is desirable to select a column size based on a low-to-moderate percentage of reinforcement so that reinforcement can be added for the final design as necessary to satisfy final column moments (including slenderness effects) without the need to change column size. Final design in the range of 1% to 2% reinforcement will result in maximum economy.

The column sizing chart presented in Fig. 5-4 is based on ACI Eq. (10-2) for tied columns loaded at an eccentricity $\simeq$ 0.10h. Column capacities for square column sizes from 8 in. to 24 in. with symmetrical reinforcement are presented. For other column sizes and shapes, and material strengths, the designer can easily develop similar column sizing charts based on ACI Eq. (10-1) or Eq. (10-2).

The simple chart will provide quick estimates for a column size required to support a factored load P_u within any desired range of reinforcement percentage. Using a total tributory factored load P_u for the lowest story of a multistory column stack, a column size can be selected with a low percentage of reinforcement. This will allow some leeway to increase steel percentage for the final design.

Suggested procedure for use of the chart to establish a column size for a typical interior column stack of Bldg. #2 is illustrated in Section 5.6.1.

5.5 - COLUMN SLENDERNESS CONSIDERATIONS

5.5.1 - Braced vs Unbraced Columns

Because of the difference not only in behavior but also in design complexity between a braced and an unbraced frame, it is necessary to establish whether

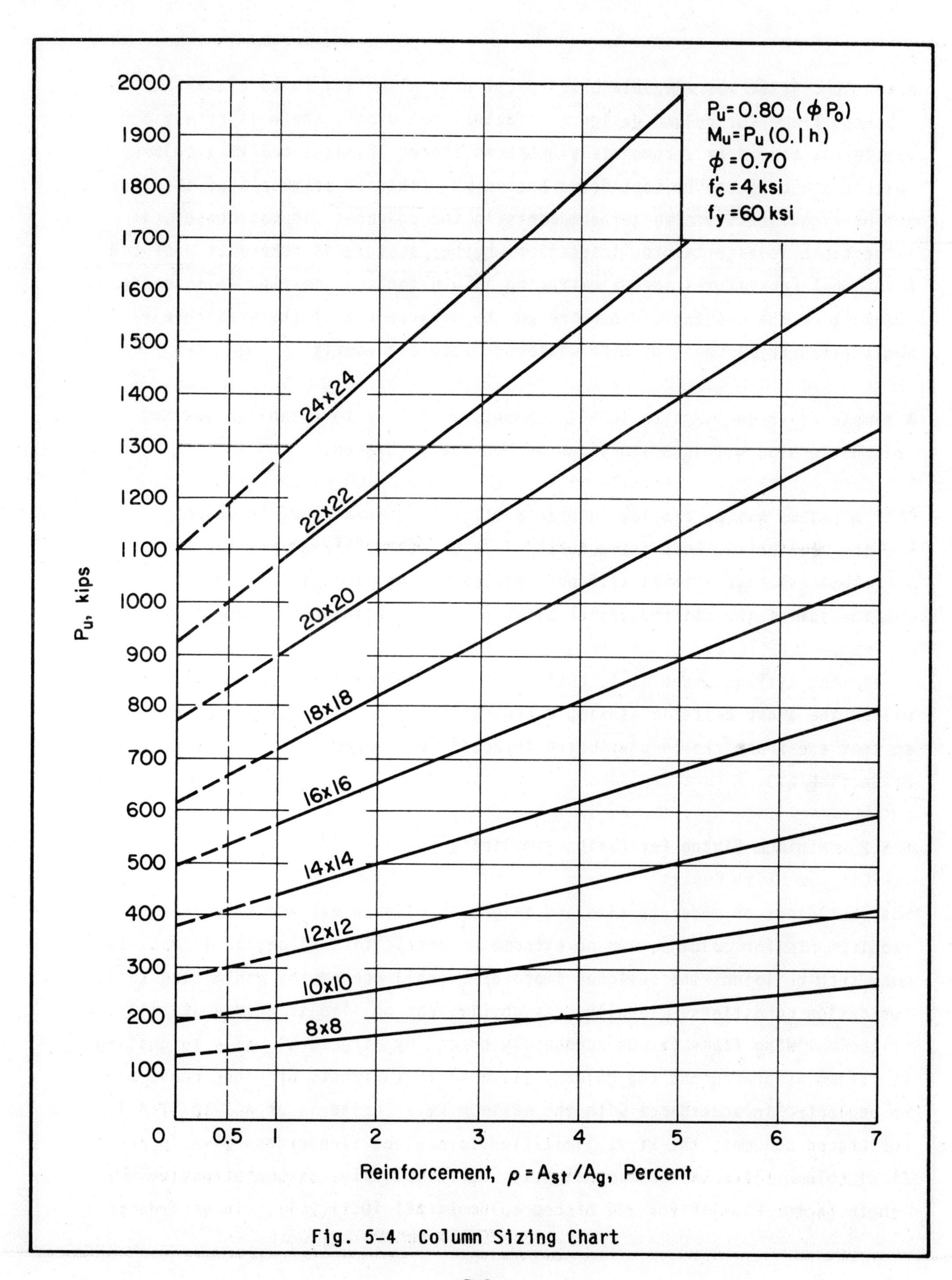

Fig. 5-4 Column Sizing Chart

a framing system has adequate bracing (structural walls) to be classified as a braced frame for column design. In actual buildings, there is rarely a completely braced or a completely unbraced frame. Realistically, a column (within a story) can be considered braced when lateral movements of the story do not significantly affect the moments in the columns. In many cases, it is possible to ascertain by inspection whether a story is braced or unbraced. Also, what constitutes adequate bracing in a given case must be left to the judgment of the designer, depending on the arrangement of the bracing elements (structural walls or other lateral bracing elements).

A simple criteria is given in ACI Commentary 10.11.2 to establish whether columns located within a story can be considered braced:

> "A column may be considered braced if located in a story in which the bracing elements (structural walls or other lateral bracing elements) have a total stiffness at least equal to six (6) times the sum of the stiffnesses of all the columns within the story."

In framing systems where it is desirable and economical to include structural walls, the above criteria is simple enough for sizing the structural walls so that the frame can be considered "braced" for column design. See Chapter 6, Section 6.3.

5.5.2 - Minimum Sizing for Design Simplicity

Most buildings of moderate size and height involve no extreme slenderness requirements for columns, and no extreme eccentricities of design loads. In many such buildings the designer improves overall economy by minimizing these two design conditions. Ideally, design time can be simplified dramatically, if the building frame can be adequately braced by structural walls to qualify the frame as braced and the columns sized so that effects of slenderness may be neglected in accordance with the maximum $k\ell_u/r$ criteria of ACI 10.11.4.1. For braced columns, the $k\ell_u/r$ limitation to neglect slenderness gives practical column sizes vs. column heights. For simplicity, assume effective length factor $k = 1.0$ for all braced columns (ACI 10.11.2.1). In accordance with ACI 10.11.4.1, effects of slenderness may be neglected when "braced" columns are sized to satisfy the following:

First story	$\ell_u/h \leq 10$
Above first story	$\ell_u/h \leq 14$

where ℓ_u is the clear height between floor members and h is the column size. The above suggested limitations are based on the following design considerations. For first-story columns, the column-footing connection is assumed to provide minimum moment resistance, with first-story columns in essentially single curvature bending. All columns above the first story are assumed in double curvature bending with approximately equal end moments. Loading conditions that lead to single curvature in the columns of a regular structure are not considered realistic design conditions. If column-footing connection, including soil resistance, is designed for moment resistance, the higher ℓ_u/h limitation of 14 would be more appropriate. For braced columns, effects of slenderness may be neglected if clear heights between floor members are not greater than the values of Table 5-2 for a given column size.

Table 5-2 Story Heights to Neglect Slenderness - Braced Columns

Column size h	Maximum clear height - ft	
	$\ell_u/h = 10$	$\ell_u/h = 14$
8	6.67	9.33
10	8.33	11.67
12	10.0	14.0
14	11.67	16.33
16	13.33	18.67
18	15.0	21.0
20	16.67	23.33
22	18.33	25.67
24	20.0	28.0

For unbraced frames, the $k\ell_u/r$ limitation to neglect column slenderness (ACI 10.11.4.2) results in unacceptable column sizes vs. column heights for most practical building framing systems. Assuming an effective length factor k = 1.2 for unbraced columns (ACI 10.11.2.2), effects of slenderness could be neglected when "unbraced" columns are sized to satisfy the following:

All columns . . . . $\ell_u/h \le 6$

Accordingly, clear story heights must not be greater than the values of Table 5-3 to eliminate the need to consider slenderness in the design of unbraced columns.

Table 5-3 Story Heights to Neglect Slenderness - Unbraced Columns

Column size h	Maximum clear height - ft $\ell_u/h = 6$
8	4.0
10	5.0
12	6.0
14	7.0
16	8.0
18	9.0
20	10.0
22	11.0
24	12.0

For some occupancies and framing schemes, these clear heights vs. minimum column size will be acceptable and slenderness effects may be neglected in design of the columns. If, however, for architectural or other reasons, the columns must be made more slender, then slenderness effects must be included in the column design.

5.5.3 - Evaluation of Column Slenderness - A Simplified Approach

Since the 1971 Code introduction of the moment magnifier technique to account for column slenderness, designers have voiced varying degrees of concern for the voluminous computations and time-consuming effort required to consider the many variables and design conditions necessary, even for the most symmetrical of building framing layouts, frame proportions, member stiffnesses, and loading conditions. Too often, "exact" procedures for application of the code provisions to account for column slenderness used more of the engineer's time than results justified. And, to further compound the problem, designers discovered that, for modest height buildings, the moment magnifier technique invariably gave seemingly unreasonable results when compared to earlier designs with comparable building types. Results of the magnifier method were deemed too conservative, and more sophisticated second-order frame analyses had to be used to give more reasonable results.

For frames without structural walls, the following discussion will attempt to isolate the many variables and design conditions that need to be considered to account for the influence of "secondary" moments in the design of unbraced columns. A simplified approach will be presented that can be considered accurate enough for buildings of modest height with regular framing layouts.

In essence, the moment magnifier method presented in ACI 10.11 is an attempt to account for a multitude of conditions producing secondary effects not accounted for by a conventional elastic frame analysis. Proper application of the "approximate" method requires considerable judgment of the many variables and design conditions. The following simplified approach is intended to reflect the intent of the magnifier method with certain simplifications in the analysis and evaluation procedure to account for slenderness effects of the columns of an unbraced framing system. The simplifications are intended to provide results accurate enough for buildings of modest height with regular framing layouts. For tall, slender, more flexible buildings and buildings with irregular proportions and member stiffnesses, modifications in the simplified procedure may need to be made, or a more exact procedure adopted. By necessity, the decision is left to the discretion of the designer.

The simplified approach is based on the concept of using an average moment magnification factor for each story of an unbraced frame. The concept is based in part on a simplified design aid published by CRSI.[5.6] Using the basic δ-magnification expression, $\delta = 1/(1 - P_u/\varphi P_c)$, an average story moment magnification factor δ_a can be derived as a simple function of the known quantity $(k\ell_u/h)$. For our selected materials f'_c = 4000 psi and f_y = 60,000 psi, and, considering only square tied columns, a critical load ratio $(P_u/\varphi P_c)$ for unbraced columns can be evaluated as follows:

$$P_c = \frac{\pi^2 EI}{(k\ell_u)^2} = \frac{2965\ (0.2 + 9.65\ \rho_t)h^2}{(k\ell_u/h)^2}$$

where EI[5.7] $= (0.2 + 1.2\ \rho_t\ E_s/E_c)\ E_c I_g$

E_s = 29000 ksi

$E_c = 57\ \sqrt{f'_c} = 3605$ ksi

$I_g = h^4/12$

Set P_u equal to maximum design axial load strength permitted by ACI Eq. (10.2):

$$P_u = \varphi P_{n(max)} = 0.80\varphi\ [0.85f'_c(A_g - A_{st}) + f_y A_{st}] = (1.9 + 31.7\rho_t)h^2$$

where $\rho_t = A_{st}/h^2$

φ = 0.70

Critical load ratio $(P_u/\varphi P_c)$ reduces to:

$$\frac{P_u}{\varphi P_c} = \left[\frac{(1.9 + 31.7\rho_t)}{2075\ (0.2 + 9.65\rho_t)}\right]\left(\frac{k\ell_u}{h}\right)^2 = \beta_s \times 10^{-3}\left(\frac{k\ell_u}{h}\right)^2$$

With P_u at $\varphi P_{n(max)}$, an average story δ_a - magnification can be evaluated as a simple function of $(k\ell_u/h)$:

$$\delta_a = \frac{1}{1 - \beta_s \times 10^{-3}\left(\frac{k\ell_u}{h}\right)^2}$$

Values of $\beta_s \times 10^{-3}$ are given in Table 5-4 for $P_u = 0.80(\varphi P_o)$. Since δ is quite sensitive to changes in the factored axial load P_u supported by a column, the tabulated values of β_s should be multiplied by the actual load ratio ($P_u/0.80\varphi P_o$) to reflect a lesser axial load than the maximum permitted. Alternatively, the designer may prefer the δ_a-magnification expressed directly as a function of the factored axial load P_u:

$$\delta_a = \frac{1}{1 - \left[\frac{P_u}{(415 + 200\,\rho_t)h^2}\right]\left(\frac{k\ell_u}{h}\right)^2}$$

where P_u is in kips and ρ_t is in percent.....$\rho_t = 100A_{st}/h^2$.

For simplification, all unbraced columns within a story may be designed using a single story δ_a-magnification evaluated for a typical interior column (highest P_u).

The δ_a-magnification applies also to columns of braced frames where slenderness must be considered. For simplicity, assume an effective length factor k = 1.0 for all braced columns. Also, the C_m factor, taken as 1.0 in the δ_a derivation, applies conservatively for braced columns.

Table 5-4 - Column Slenderness Factor $\beta_s \times 10^{-3}$ for $P_u = 0.80\ (\varphi P_o)$

ρ_t	.000	.001	.002	.003	.004	.005	.006	.007	.008	.009
0.00						4.00	3.91	3.82	3.74	3.67
0.01	3.60	3.54	3.48	3.42	3.37	3.32	3.27	3.23	3.19	3.15
0.02	3.11	3.07	3.04	3.00	2.97	2.94	2.91	2.88	2.86	2.83
0.03	2.81	2.78	2.76	2.74	2.72	2.70	2.68	2.66	2.64	2.62
0.04	2.61	2.59	2.57	2.56	2.54	2.53	2.51	2.50	2.49	2.47
0.05	2.46	2.45	2.44	2.43	2.41	2.40	2.39	2.38	2.37	2.36
0.06	2.35	2.34	2.33	2.32	2.32	2.31	2.30	2.29	2.29	2.27
0.07	2.27	2.26	2.25	2.25	2.24	2.23	2.22	2.22	2.21	2.21
0.08	2.20									

5.5.4 - Summary of Simplifications for Slenderness Considerations

The following simplifications in loading and slenderness evaluation are suggested. For buildings of modest height and regular framing layouts, the suggested shortcuts will alter the final results very little.

(1) For braced frames, size all columns to satisfy ℓ_u/h limitation to neglect slenderness effects. If slenderness must be considered, use an effective length factor k equal to 1.0 for all columns.

(2) For unbraced frames, use an effective length factor k equal to 1.2 for all columns.

(3) For unbraced frames, compute an average story δ-magnification for a typical interior column (highest P_u) and use same δ_a for all columns within a story.

5.6 - PROCEDURE FOR SIMPLIFIED COLUMN DESIGN

The following procedure is suggested for design of a multistory column stack using the simplifications and column design charts presented in this chapter. The step-by-step procedure addresses specifically columns of unbraced frames with the frame designed to resist both gravity loads (dead + live) and lateral loads (wind). For braced frames with structural walls resisting the lateral loads and the columns sized within the ℓ/h limitation to neglect slenderness, the design reduces to consideration of gravity loads only and short column capacities for bar selection.

STEP (1) - LOAD DATA

(a) Gravity Loads

Determine total of tributary factored loads P_u for each floor level of column stack being considered. Consider only P_u for full gravity loads (dead + live) on all floors (including live load reduction) and roof. Include 4 kips per floor for column weight. Determine column moments due to gravity loads. For interior columns (unless a general analysis is made to evaluate gravity load moments from alternate span loading) compute maximum column moments by ACI Eq. (13-4); see Chapter 4, Section 4.5.

(b) Lateral Loads
Gather column end moments from wind load analysis for column stack being considered.

STEP (2) - LOAD COMBINATIONS

For gravity (dead + live) plus wind loading, ACI 9.2.2 specifies three load combinations that need to be considered:

gravity loads:

$$U = 1.4D + 1.7L \qquad \text{Eq. (9-1)}$$

gravity plus wind loads:

$$U = 0.75(1.4D + 1.7L + 1.7W) \qquad \text{Eq. (9-2)}$$

or

$$U = 0.9D + 1.3W \qquad \text{Eq. (9-3)}$$

Generally, for unbraced columns subjected to wind moments, Eq. (9-2) will be the controlling load combination. Once a column size is selected for the column stack (STEP 3), selecting the reinforcement for each floor level will be less of a trial procedure if the initial selection is based on Eq. (9-2). Eqs. (9-1) and (9-3) will usually absorb more of the designer's time than results justify.

STEP (3) - COLUMN SIZE

For the multistory column stack, size columns beginning with the most heavily loaded (first story) and use the same column size for full height of building...with lesser amounts of reinforcement in the upper stories. For largest P_u (first story) select a column size from Fig. 5-4 with slightly less than 1% reinforcement.

Ideally, a column size with reinforcement in the range of 1% to 2% for a final design will result in maximum economy. Depending on total number of stories, difference in story heights, and magnitude of wind loads, 4% to 6% reinforcement may be required in the first story columns, with decreasing amounts in the upper stories for overall cost economy. Note: The percentage of reinforcement should not exceed 4% if the column bars are to be lap spliced.

STEP (4) - SELECTION OF REINFORCEMENT

(a) Reinforcing bars for each floor level beginning with the first story columns are selected directly from the column design charts, Figs. 5-6 through 5-14, for the respective load combinations.

(b) To include effects of column slenderness, a somewhat higher reinforcement percentage (say, two bar arrangements above the amount adequate for a short column) is selected as a first trial. The trial selection is then checked as a slender column with the moment M_u magnified by the δ_a-magnification to account for secondary moments due to column slenderness.

For all exterior columns within a given story, use the same δ_a-magnification as used for the final design of the interior columns of the same story.

After the reinforcing bars are set for the first story, bar selection is repeated for succeeding floors for the respective load combinations to complete the design of a column stack.

5.7 - EXAMPLES: SIMPLIFIED DESIGN FOR COLUMNS

The following three examples illustrate use of the simplified design data presented in Chapter 5 for design of columns.

5.7.1 - Example: Design an Interior Column Stack for Bldg. #2

Alternate (1) - Slab and Column Framing Without Structural Walls (~~Unbalanced~~ Unbraced Frame).

f'_c = 4000 psi

f_y = 60,000 psi

Column size and selection of reinforcing bars for the interior column stack follows the procedure outlined in Section 5.6.

(1) - LOAD DATA

Roof: LL = 20 psf   Floors: LL = 50 psf

DL = 122 psf   DL = 142 psf (9 in. slab)

Calculations for the 1st-story column are as follows:

(a) Total factored load

roof: P_u = 1.4(0.122 x 24 x 20) + 1.7(0.02 x 24 x 20) = 98^k

Reduced live load for a total influence area of 4 floors

A_I = 24 x 20 x 4 x 4 = 7680 sq ft

$L_r = 50(0.25 + 15/\sqrt{7680})$ = 21 psf

4 floors:

P_u = 1.4(0.142 x 24 x 20 x 4) + 1.7(0.021 x 24 x 20 x 4) = 450^k

column wt = 1.4(4x5) = 28^k

ΣP_u = 576^k

(b) Factored moments

gravity loads: $M_u = 0.035 w_\ell \ell_2 \ell_n^2 = 0.035(0.05 \times 1.7) 24 \times 18.67^2$ = 25 'k

portion of M_u to 1st-story column $25(\frac{12}{12+15})$ = 11 'k

wind loads: from wind load analysis M_u = 1.7(94.7) = 161 'k

(See Fig. 2-14)

(2) - LOAD COMBINATIONS

For the 1st-story column:

gravity loads: $P_u = 576^k$ Eq. (9-1)

$M_u = 11'^k$

gravity + wind loads: $P_u = 0.75(576) = 432^k$ Eq. (9-2)

$M_u = 0.75(11 + 161) = 129'^k$

or

$P_u = 0.9(59 + 273 + 20) = 316^k$ Eq. (9-3)

$M_u = 1.3(94.7) = 123'^k$

Factored loads and moments, and load combinations, for the 2nd- through 5th-story columns are calculated in the same manner. The various load combinations to be considered for the 5-story interior-column stack are summarized as follows:

Bldg. #2 Alternate (1) Interior Column Stack	Gravity Loads		Gravity + Wind Loads			
	Eq. (9-1)		Eq. (9-2)		Eq. (9-3)	
	P_u	M_u	P_u	M_u	P_u	M_u
5th	104	10*	78	19	57	12
4th	229	13	172	44	122	35
3rd	346	13	260	65	187	57
2nd	462	14	347	86	252	77
1st	576	11	432	129	316	123

*$M_u = 0.035\ (0.02 \times 1.7)\ 24 \times 18.67^2 = 10'^k$

(3) - COLUMN SIZE

For largest $P_u = 576^k$, select a 16x16 column with slightly less than 1% reinforcement from column sizing chart, Fig. 5-4. Use same column size for full height of building and vary column reinforcing for each story.

Column slenderness consideration (unbraced frame):

1st-story $\ell_u/h = (15x12)/16 = 11.3$

2nd- through 5th-story $\ell_u/h = (12x12)/16 = 9.0 \quad > 6$

Also from Table 5-3, with h = 16: $\ell_u = 8.0$ ft < 12 ft

Column slenderness must be considered

For slenderness evaluation:

1st-story $k\ell_u/h = 1.2(15x12)/16 = 13.5$

2nd- through 5th-story $k\ell_u/h = 1.2(12x12)/16 = 10.8$

(4) - SELECTION OF REINFORCEMENT

(a) For the 1st-story column:

Using the 16x16 column design chart, Fig. 5-10:

For gravity loads: $P_u = 576^k$ 4 #8 bars req'd ($\rho_t = 1.23\%$)

$M_u = 11^{'k}$

For gravity + wind: $P_u = 432^k$ 4 #11 bars req'd ($\rho_t = 2.44\%$)

$M_u = 129^{'k}$

or

$P_u = 316^k$ 4 #8 bars req'd ($\rho_t = 1.23\%$)

$M_u = 123^{'k}$

Eq. (9-2) is the controlling load combination by a significant margin due to high wind moments. Final bar selection, including slenderness effects, will be based on Eq. (9-2). As a first trial, select 8 #10 bar arrangement.

For 8 #10 bars: $\rho_t = 3.97\%$

$0.8\varphi P_o = 809^k$

Compute column slenderness factor for $P_u = 432^k$ from Table 5-4:

$$\beta_s = 2.61(432/809) = 1.39 \times 10^{-3}$$

$$\delta_a = \frac{1}{1 - 0.00139(13.5)^2} = 1.34$$

or if preferred:

$$\delta_a = \frac{1}{1 - \left[\frac{432}{(415 + 200 \times 3.97)16^2}\right](13.5)^2} = 1.34$$

$$P_u = 432^k$$
$$M_u = 129(1.34) = 173^{'k}$$

For the new load combination including slenderness effects, the 8 #10 bars are adequate.

(b) For the 2nd-story column:

$P_u = 462^k$ 4 #7 bars req'd ($\rho_t = 0.94\%$)

$M_u = 14^{'k}$

$P_u = 347^k$ 4 #7 bars req'd

$M_u = 86^{'k}$

$P_u = 252^k$ 4 #6 bars req'd ($\rho_t = 0.69\%$)

$M_u = 79^{'k}$

Try 4 #8 bars: $\rho_t = 1.23\%$

$0.8\varphi P_o = 588^k$

From Table 5-4: $\beta_s = 3.46(347/588) = 2.04 \times 10^{-3}$

$$\delta_a = \frac{1}{1 - 0.00204(10.8)^2} = 1.31$$

$P_u = 347^k$

$M_u = 86(1.31) = 113^{'k}$ <u>4 #8 bars</u> OK

Bar selection for the 3rd- through 5th-story columns is repeated in the same manner. Results are as follows:

3rd-story - <u>Use 4 #7 bars</u>

4th- and 5th - <u>Use 4 #6 bars</u>

5.7.2 - Example: Design Interior Column Stack for Bldg. #2 Alternate (2) - Slab and Column Framing with Structural Walls (Braced Frame). } p1-13

$f'_c = 4000$ psi

$f_y = 60{,}000$ psi

For the Alternate (2) framing, columns are designed for gravity loading only; the structural walls are designed to resist total wind loading.

(1) - LOAD DATA

Roof: LL = 20 psf Floors: LL = 50 psf

DL = 122 psf DL = 136 psf (8-1/2 in. slab) 9"

Calculations for 1st-story column are as follows:

(a) Total factored load

roof: $P_u = 1.4(0.122 \times 24 \times 20) + 1.7(0.02 \times 24 \times 20) = 98^k$

live load reduction:

A_I (4 panels, 4 floors) = 24 x 20 x 4 x 4 = 7680 sq ft

$L_r = 50(0.25 + 15/\sqrt{7680}) = 21$ psf

SBC: R = 4x24x20x0.0008 = 154% > 60% max

L_r = 50x0.40 = 20psf

4 floors:

$$P_u = 1.4(0.136 \times 24 \times 20 \times 4) + 1.7(0.021 \times 24 \times 20 \times 4) = 434^k$$

$$\text{Column wt} = 1.4(4\text{x}5) = 28$$

$$\Sigma P_u = 560^k$$

(b) Factored gravity load moment

From Section 4.5: only if adjacent & transverse spans equal

$$M_u = 0.035 w_\ell \ell_2 \ell_n^2 = 0.035(0.05 \times 1.7)24 \times 18.67^2 = 25^{'k}$$

portion to 1st-story column $M_u = 25(12/27) = 11^{'k}$ ← 1st + 2nd story

Similar calculations for each succeeding floor yield the following results:

Bldg. #2 Alternate (2) Int. Col. Stack	Eq. (9-1)	
	P_u	M_u
5th	104	10
4th	224	13
3rd	338	13
2nd	449	14
1st	560	11

(2) - LOAD COMBINATIONS

Only ACI Eq. (9-1) need be considered for gravity loads as summerized above.

(3) - COLUMN SIZE

For largest $P_u = 560^k$, select a 16x16 column with slightly less than 1% reinforcement from column sizing chart, Fig. 5-4. (p5-9) Use same column size for full height of building and vary reinforcing for each story.

Column slenderness consideration (braced frame):

(p5-11)
From Table 5-2, with h = 16:

1st-story ℓ_u = 13.33 ft < 15 ft

2nd- through 5th-story ℓ_u = 18.67 ft > 12 ft

Slenderness as a braced column needs to be considered for the 1st-story columns only. For slenderness evaluation:

$k\ell_u/h = 1.0(15x12)/16 = 11.3$

(4) - SELECTION OF REINFORCEMENT

(a) For 1st-story column: Fig 5-10 (p5-39)

$P_u = 560^k$ 4 #8 bars req'd (ρ_t = 1.23%)

$M_u = 11'^k$

With small moment, check slenderness effects for 4 #8 bar arrangement.

For 4 #8 bars: $\rho_t = 1.23\%$

$0.8\varphi P_o = 588^k$ (p 5-39)

(p 5-15)
From Table 5-4: $\beta_s = 3.46(560/588) = 3.3 \times 10^{-3}$

$$\delta_a = \frac{1}{1 - 0.0033(11.3)^2} = 1.73$$

$P_u = 560^k$ <u>4 #8 bars</u> OK (p 5-39)

$M_u = 11(1.73) = 19'^k$

(b) 2nd- through 5th-story columns (slenderness need not be considered): Referring to Fig. 5-10:

2nd - $P_u = 449^k$ Use 4 #7 bars (ρ_t = 0.94%)

$M_u = 14'^k$

3rd - $P_u = 338^k$ Use 4 #6 bars (ρ_t = 0.69%)

$M_u = 13'^k$

4th & 5th - Use 4 #6 bars (minimum for 16x16 column)

Note: If, for architectural considerations, a smaller column size is preferred, a 14x14 column would require the following reinforcement. Using Fig. 5-9 for bar selection:

14x14 Column Alternative

1st-story - 4 #11 bars (ρ_t = 3.18%)
2nd-story - 4 #7 bars (ρ_t = 1.22%)
3rd-story - 4 #6 bars (ρ_t = 0.89%)
4th & 5th - 4 #5 bars (ρ_t = 0.63%)

5.7.3 - Example: Design Exterior Column Stack (EW-Column Line) for Bldg. #1 - 3-Story Pan Joist Construction

f'_c = 4000 psi

f_y = 60,000 psi

Bldg. #1 is an unbraced frame with columns designed for both gravity and wind loads.

(1) - LOAD DATA

Roof: LL = 12 psf   Floors: LL = 60 psf

DL = 105 psf   DL = 145 psf

Calculations for 1st-story column are as follows:

(a) Total factored load supported by 1st-story edge columns

roof: $P_u = 1.4(0.105 \times 30 \times 15) + 1.7(0.012 \times 30 \times 15) = 75^k$

live load reduction:

A_I (2 panels, 2 floors) = 30 x 30 x 2 x 2 = 3600 sq ft

See Table 2-1: $L_r = 60(0.50) = 30$ psf

2 floors:

$$P_u = 1.4(0.145 \times 30 \times 15 \times 2) + 1.7(0.030 \times 30 \times 15 \times 2) = 229^k$$

$$\text{Column wt} = 1.4(4 \times 3) = \underline{17^k}$$

$$\Sigma P_u = 321^k$$

(b) Factored moments in 1st-story edge columns

gravity loads: $M_u = 405.9^{'k}$

[see Section 3.8.3 - Step (2), M_u @ ext. columns]

portion of M_u to 1st-story column = 405.9/2 $= 203^{'k}$

wind loads: see Fig. 2-12, $M_u = 1.7(42.1) = 72^{'k}$

$P_u = 1.7(\pm 7.6) = \pm 13^k$

(2) - LOAD COMBINATIONS

For the 1st-story column:

gravity loads: $P_u = 321^k$ Eq. (9-1)

$M_u = 203^{'k}$

gravity + wind loads: $P_u = 0.75(321 + 13) = 251^k$ Eq. (9-2)

$M_u = 0.75(203 + 72) = 206^{'k}$

or

$P_u = 0.9(47+131+12) + 1.3(7.6) = 181^k$ Eq. (9-3)

$M_u = 1.3(42.1) = 55^{'k}$

Factored loads and moments, and load combinations, for the 2nd- and 3rd-story columns are calculated in the same manner. The various load combinations to be considered for the 3-story edge-column stack are summarized as follows:

Bldg. #1 Edge-Column Stack (EW-Column Line)	Gravity Loads		Gravity + Wind Loads			
	Eq. (9-1)		Eq. (9-2)		Eq. (9-3)	
	P_u	M_u	P_u	M_u	P_u	M_u
3rd	81	256*	62	204	47	12
2nd	206	203	158	186	113	35
1st	321	203	251	206	181	55

*$w_u = (1.4 \times 0.105 + 1.7 \times 0.012)30 = 5.02$ klf

$M_u = 5.02 \times 28.58^2/16 = 256'^k$

(3) - **COLUMN SIZE**

For edge columns, with total moment from floor system transferred directly to the edge columns, initial selection of column size is best determined by referring directly to the column design charts and select a column size based on required moment strength. For largest M_u = 256, a 16x16 column appears adequate (Fig. 5-10).

Column slenderness consideration (unbraced frame):

From Table 5-3, with h = 16:

$\ell_u = 8.0 < 13$ (slenderness must be considered).

For slenderness evaluation:

$k\ell_u/h = 1.2(13 \times 12)/16 = 11.7$

(4) - SELECTION OF REINFORCEMENT

(a) 1st-story column:

Using the 16x16 column design chart, Fig. 5-10:

For gravity loads: $P_u = 321^k$ 8 #10 bars req'd

$M_u = 203^{'k}$

For gravity + wind: $P_u = 251^k$

$M_u = 206^{'k}$

or   4 #6 bars req'd

$P_u = 181^k$

$M_u = 55^{'k}$

With large column moments from gravity loads, ACI Eq. (9-1) becomes more significant. Select final reinforcement (including slenderness effects) for gravity load combination.

Try 8 #11 bars: $\rho_t = 4.88\%$

$0.8\varphi P_o = 883^k$

From Table 5-4; for $P_u = 321^k$

$$\beta_s = 2.47(321/883) = 0.90\text{x}10^{-3}$$

$$\delta_a = \frac{1}{1 - 0.0009(11.7)^2} = 1.14^*$$

*Note: For exterior columns, the same δ_a-magnification as used for the final design of the interior columns of the same story should be used also for the exterior columns.

$P_u = 321^k$

$M_u = 203(1.14) = 231^{'k}$

Referring to Fig. 5-10, the 8 #11 bars are not adequate. By inspection, the next bar arrangement for the 16x16 column, 12 #10 bars, will have sufficient capacity including slenderness. The gravity + wind load combination will also be satisfied by the 12 #10 bars. Use 12 #10 bars for 1st-story column.

(b) 2nd-story column:

gravity load: $P_u = 206^k$ 8 #9 bars req'd
$M_u = 203'^k$

Try 8 #10 bars including slenderness: $\rho_t = 3.97\%$
$0.8\varphi P_o = 809^k$

$\beta_s = 2.61(206/809) = 0.66 \times 10^{-3}$

$$\delta_a = \frac{1}{1 - 0.00066(11.7)^2} = 1.10$$

$P_u = 206^k$
$M_u = 203(1.10) = 223'^k$ 8 #10 bars OK

(c) 3rd-story column:

gravity load: $P_u = 81^k$ 12 #10 bars req'd
$M_u = 256'^k$

Check 12 #10 bars including slenderness: $\rho_t = 5.95\%$
$0.8\varphi P_o = 970^k$

$\beta_s = 2.35(81/970) = 0.20 \times 10^{-3}$

$$\delta_a = \frac{1}{1 - 0.0002(11.7)^2} = 1.03$$

$P_u = 81^k$
$M_u = 1.03(256) = 264'^k$ 12 #10 bars OK

5.8 DESIGN FOR COLUMN SHEAR STRENGTH

When columns are required to resist the lateral forces from wind loads (unbraced frames), column shear strength needs to be checked. For square columns from 8 in. to 24 in., Fig. 5-5 can be used to readily check shear strength provided by concrete φV_c. Shear strength φV_c is based on ACI Eq. (11-4), using the largest bar size from the corresponding column design charts of Figs. 5-6 through 5-14 to evaluate φV_c. ACI Eq. (9-3) should be used to check column shear strength:

$$U = 0.9D + 1.3W$$

$$P_u = 0.9D$$

$$H_u = 1.3W$$

If H_u is greater than φV_c, spacing of column ties will need to be reduced to an effective stirrup spacing for additional shear strength φV_s. Using the three standard stirrup spacings (see Chapter 3, Section 3.6), the following values of φV_s may be used to increase column shear strength, where $H_u \leq \varphi V_c + \varphi V_s$.

Tie spacing	φV_s - #3 ties*	φV_s - #4 ties*
d/2	22 kips	40 kips
d/3	33 kips	60 kips
d/4	44 kips	80 kips

*2 legs, Grade 60 bars.

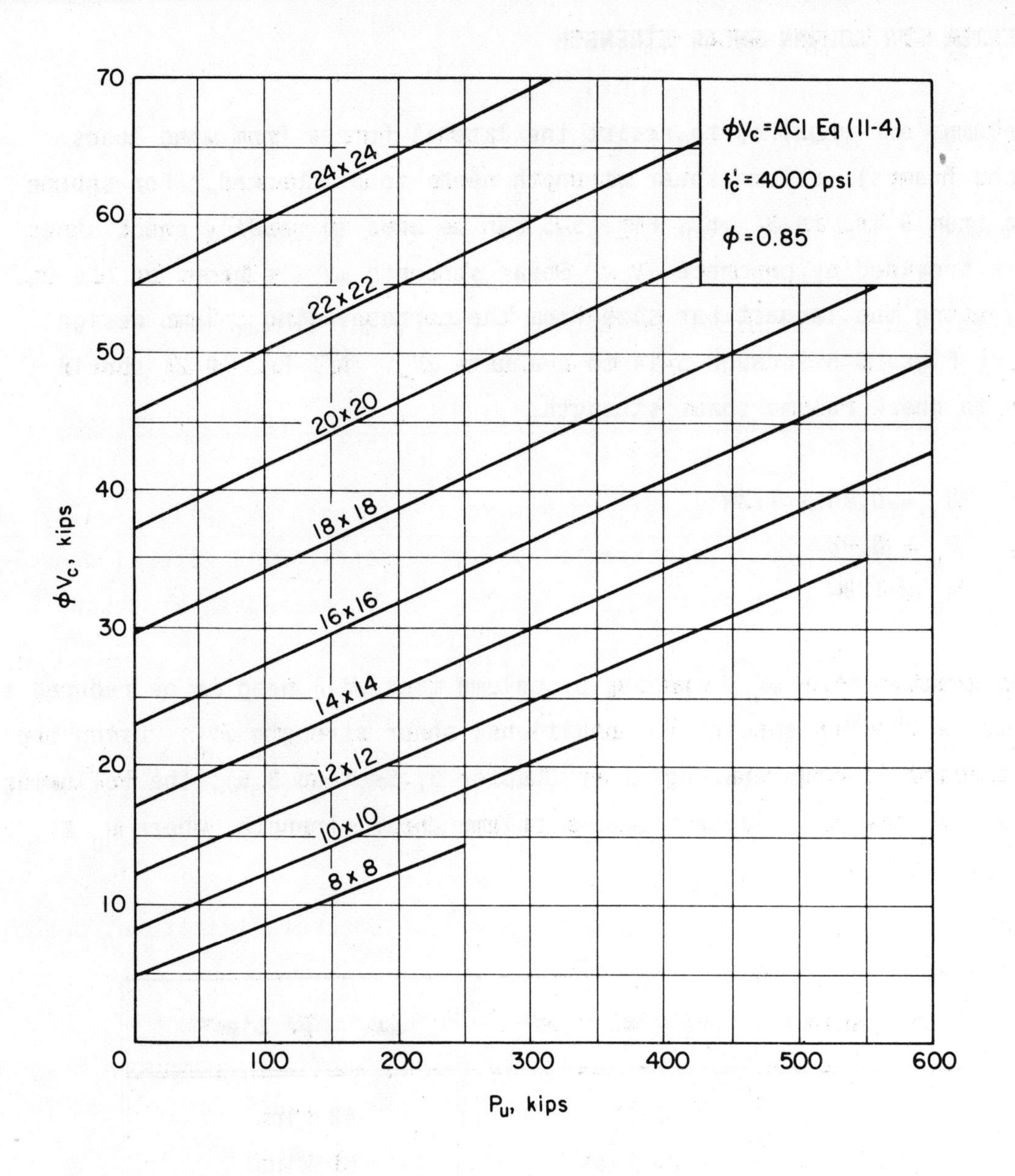

Fig. 5-5 Column Shear Strength φV_c

5.8.1 Example - Design for Column Shear Strength

Check shear strength for the interior columns of Bldg. #2 Alternate (1) - slab and column framing without structural walls. For wind in NS-direction (See Fig. 2-14), $H = 12.63^k$.

$$P_u = 0.9(59 + 273 + 20) = 316^k$$
$$H_u = 1.3(12.63) = 16.42^k$$

From Fig. 5-5, for 16 x 16 column:

$$\varphi V_c = 37^k \gg 16.42^k \quad \text{OK}$$

Column shear strength is adequate. Use column ties at regular tie spacing.

Column Ties: Using Table 5-1 and Fig. 5-2; for 8 #10 column bars:

Use #3 ties @ 16 in. with one-piece tie detail at each tie spacing.

Note: For lower height buildings, column shear strength φV_c will usually be more than adequate to resist lateral wind loads; rarely will lateral wind loads be high enough to require reduced tie spacing for increased shear strength.

Selected References

5.1 Strength Design of Reinforced Concrete Columns, Portland Cement Association, Skokie, Ill. EB009D, 49 pp.

5.2 Design Handbook, Volume 2, Columns, American Concrete Institute, Detroit, Mich. SP-17A(78), 1978, 214 pp.

5.3 Strength Design of Reinforced Concrete Column Sections, Computer Program, Portland Cement Association, Skokie, Ill. XP014D.

5.4 CRSI Handbook, Concrete Reinforcing Steel Institute, Schaumburg, Ill. 1982.

5.5 "Saving Steel in Columns," PSI - Product Services Information, Bulletin 7702A, Concrete Reinforcing Steel Institute, Schaumburg, Ill.

5.6 "Combined Strength-Slenderness One-Step Design for Columns in Ordinary Structures," PSI-Product Services Information, Bulletin 7803-A, Concrete Reinforcing Steel Institute, Schaumburg, Ill.

5.7 MacGregor, J.G., Oelhafen, V. H., and Hage, S.E., "A Re-examination of the EI Value for Slender Columns," Reinforced Concrete Columns (SP-50), American Concrete Institute, 1975, pp. 1-40.

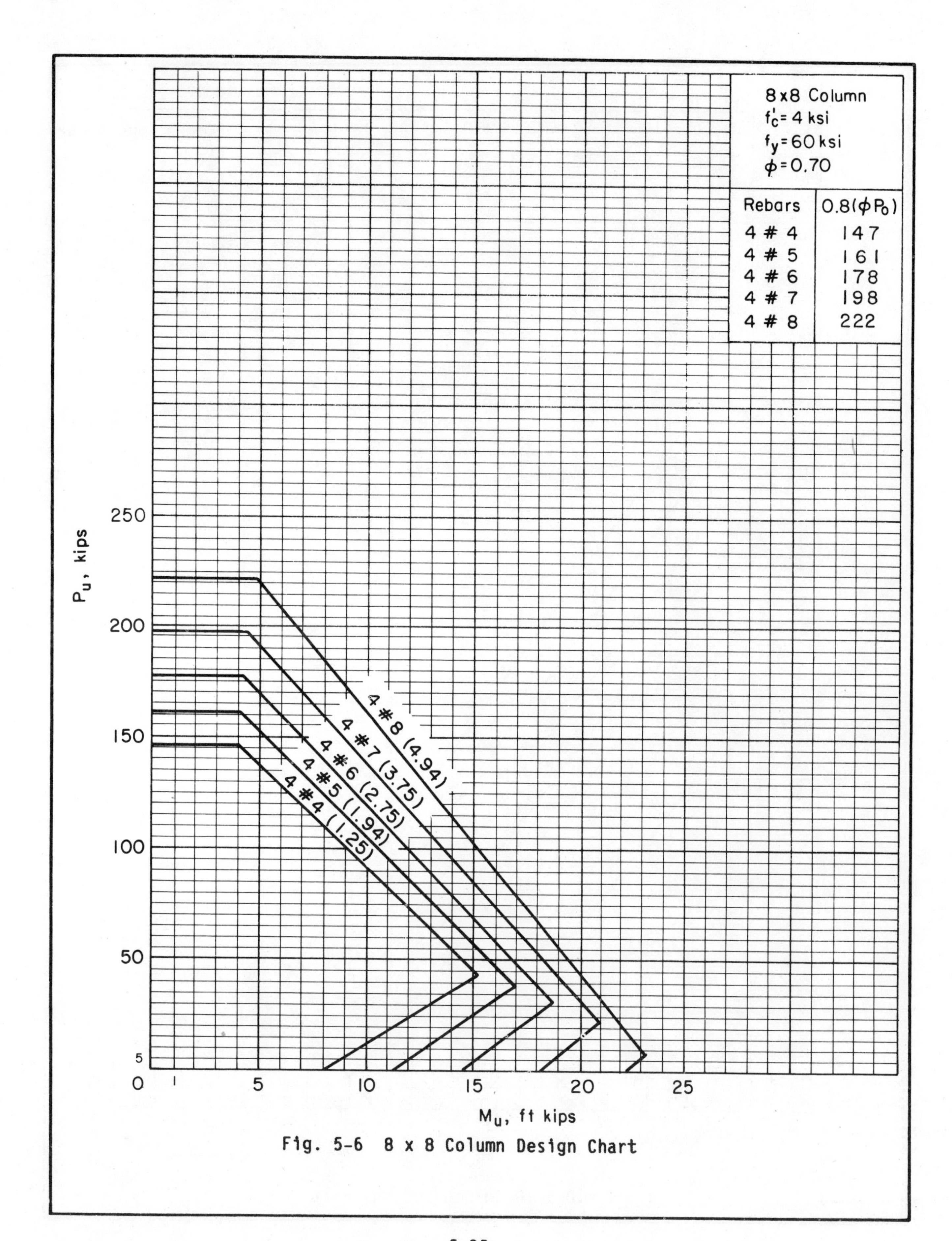

Fig. 5-6 8 x 8 Column Design Chart

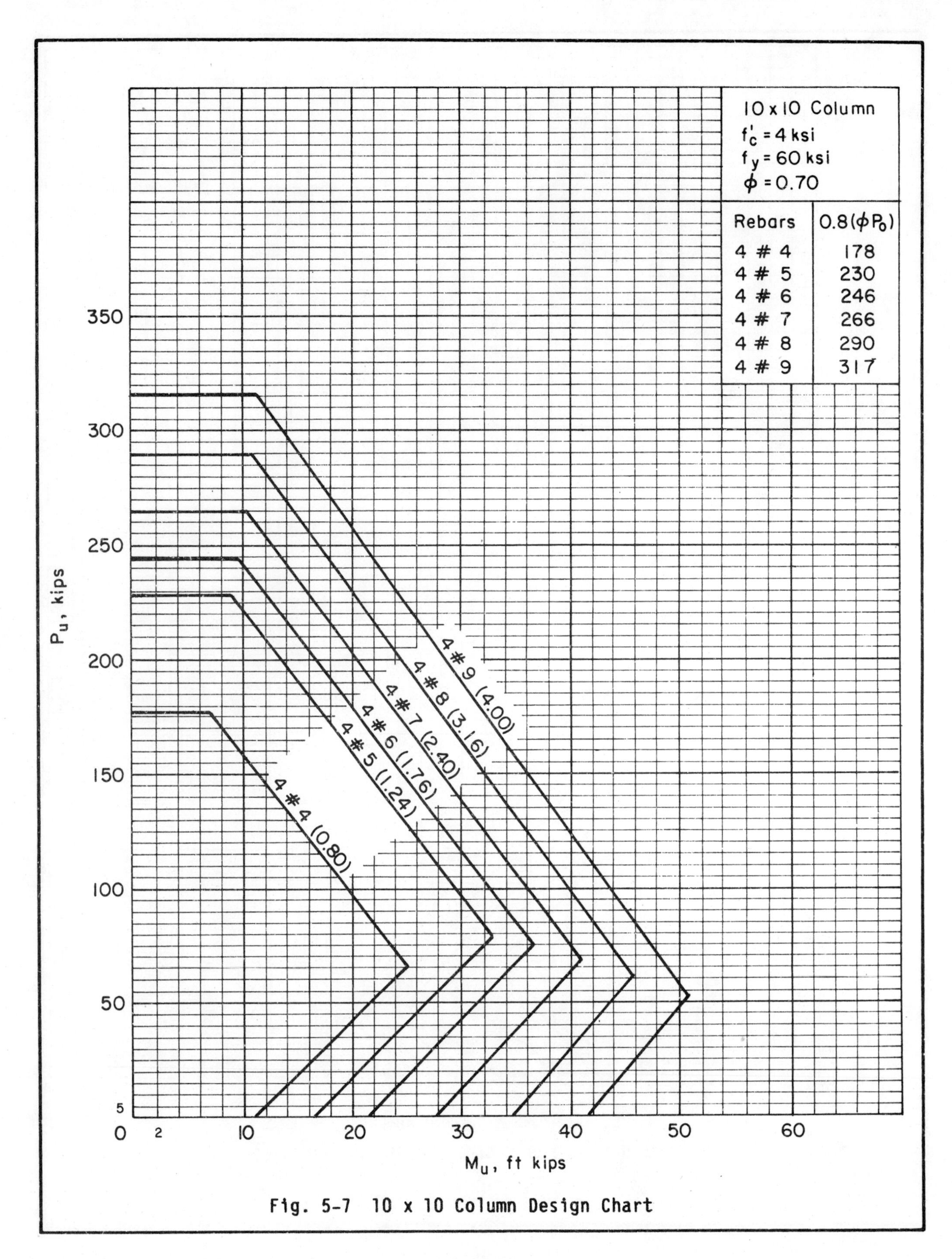

Fig. 5-7 10 x 10 Column Design Chart

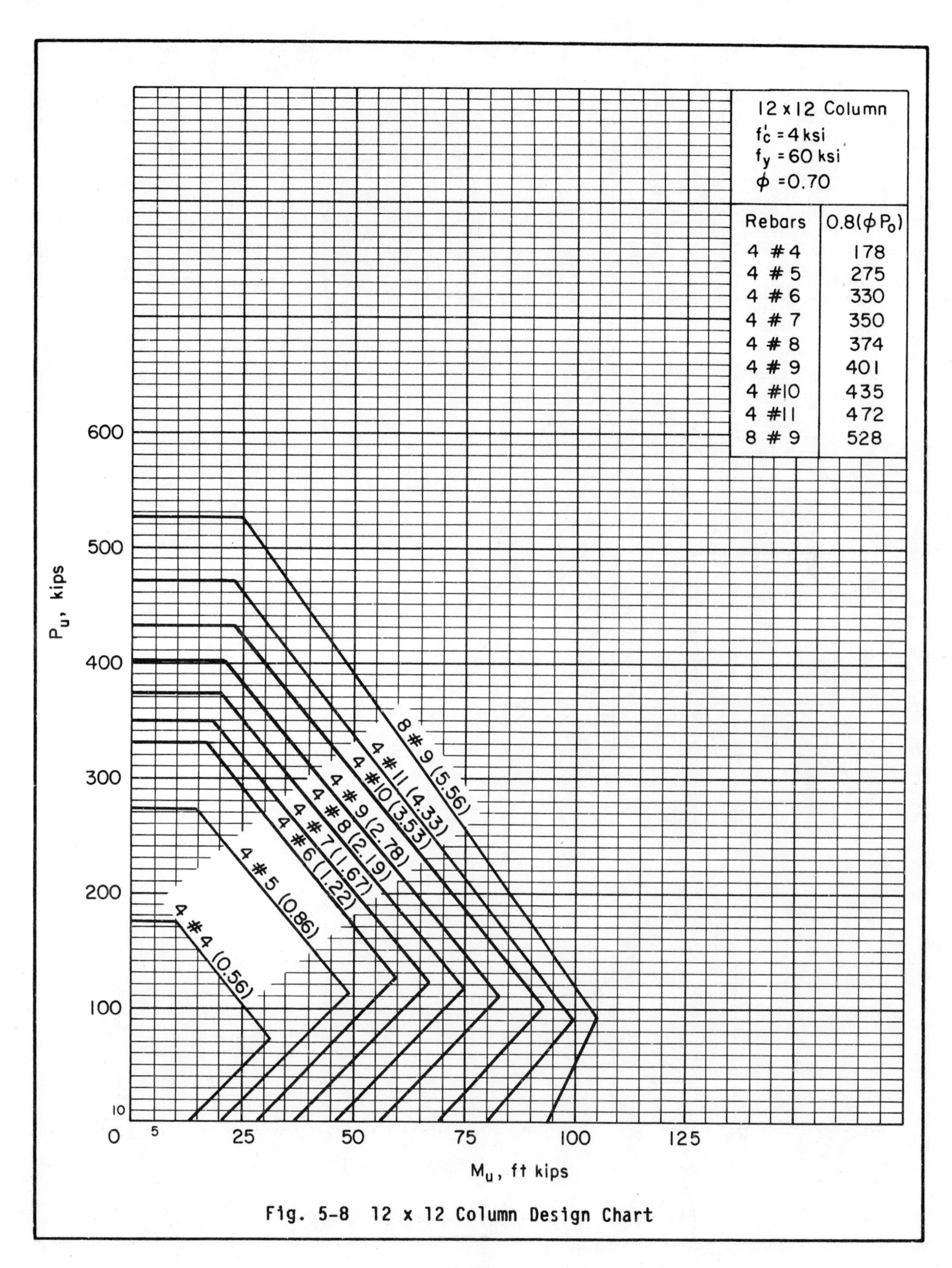

Fig. 5-8 12 x 12 Column Design Chart

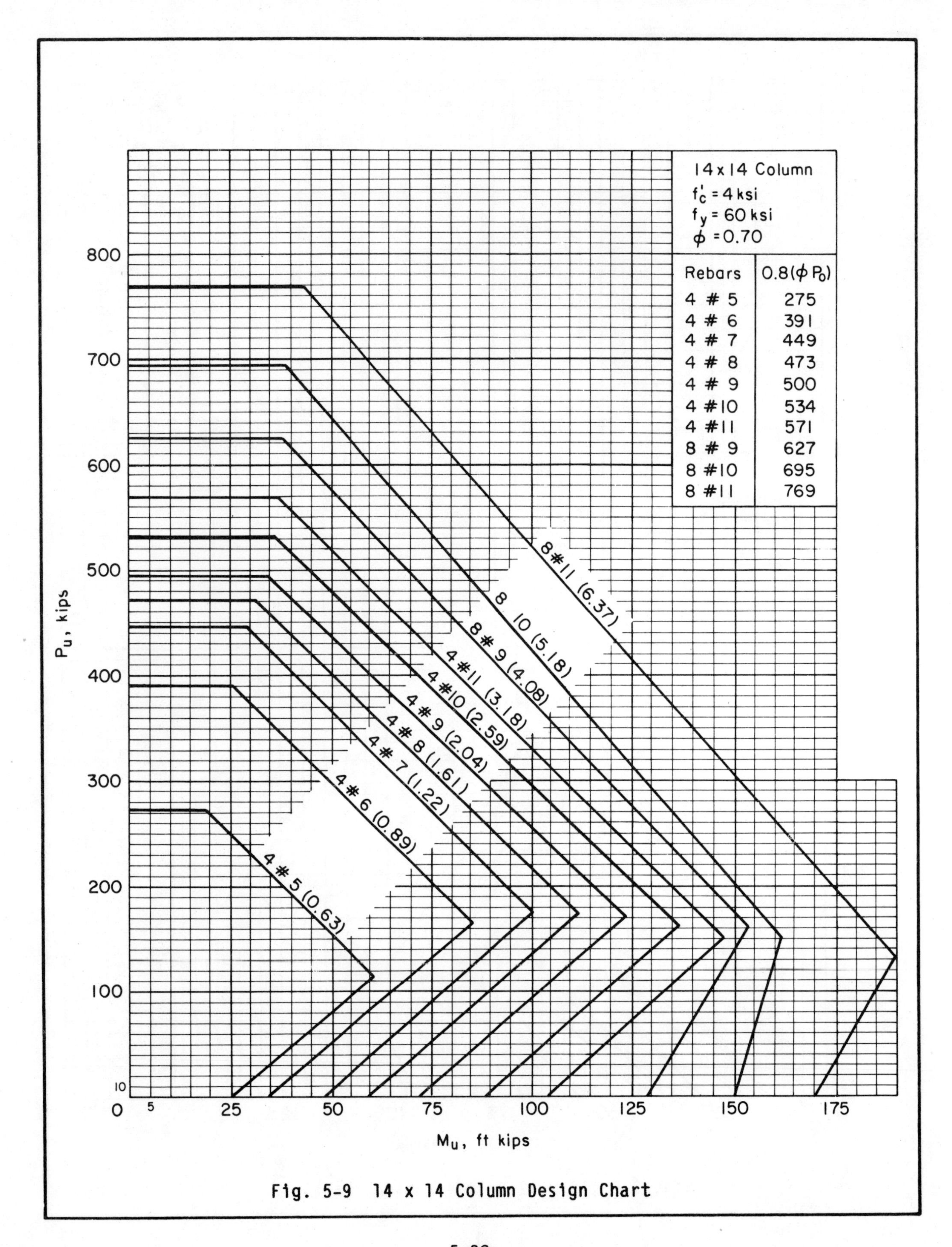

Fig. 5-9 14 x 14 Column Design Chart

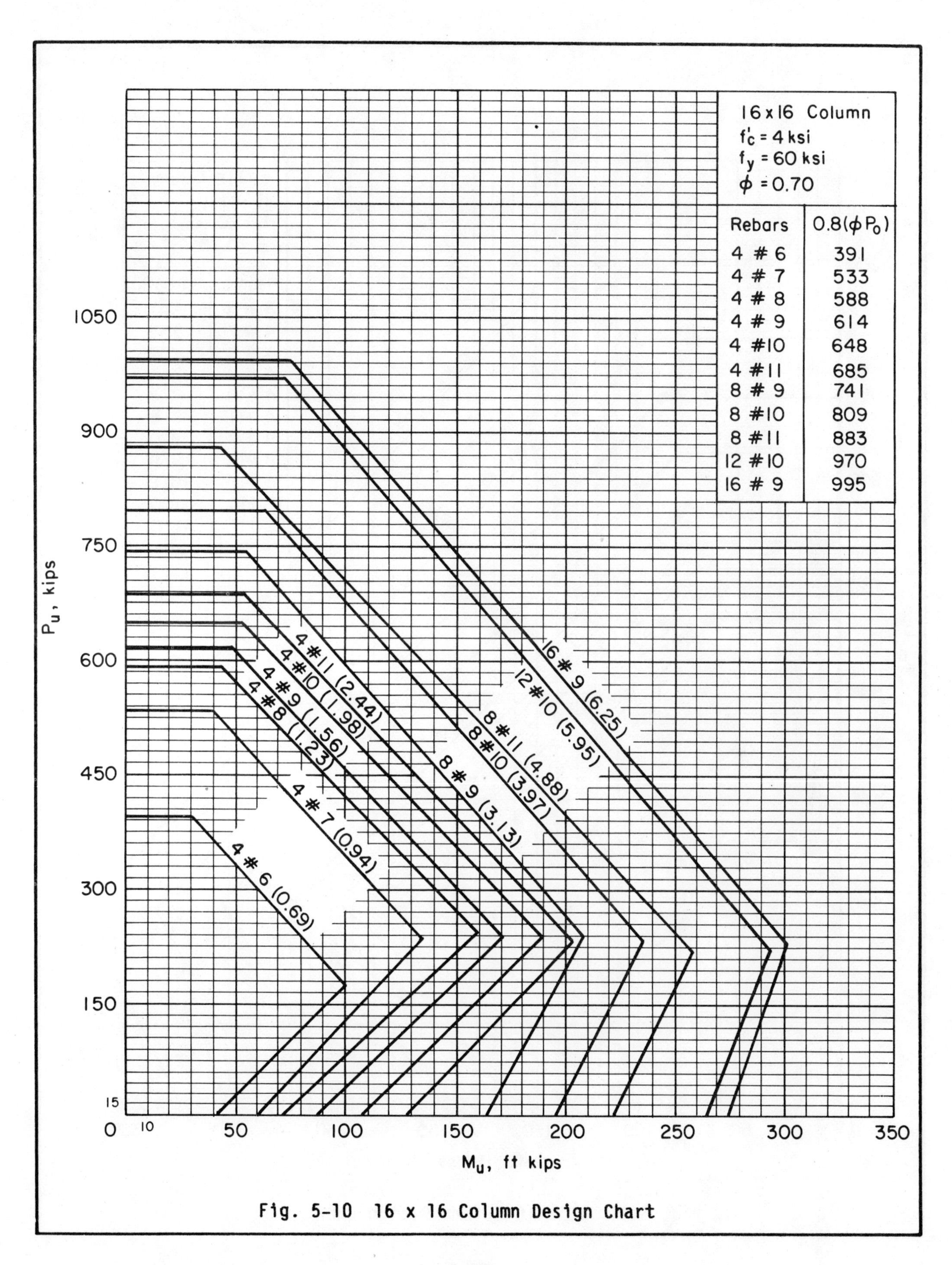

Fig. 5-10 16 x 16 Column Design Chart

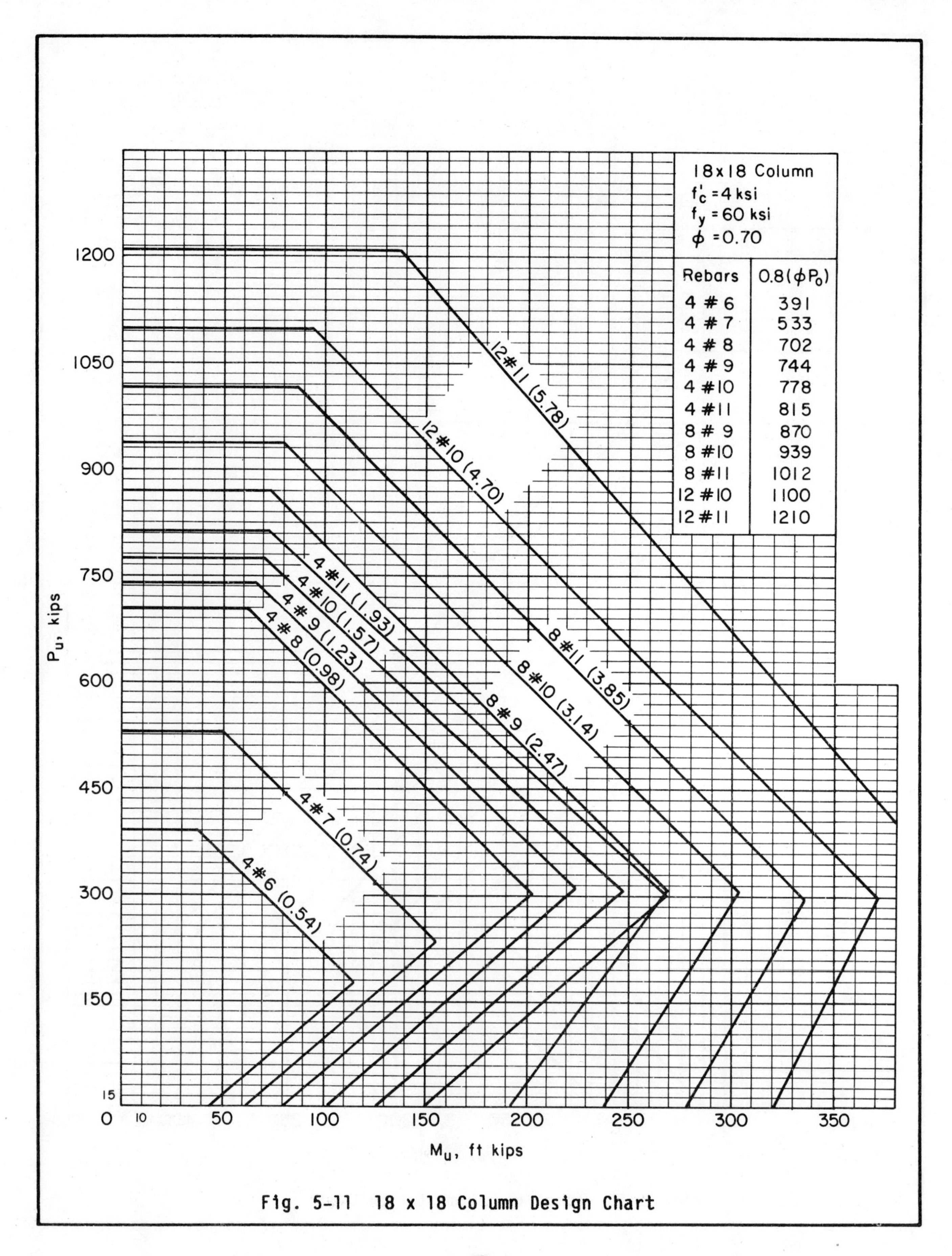

Fig. 5-11 18 x 18 Column Design Chart

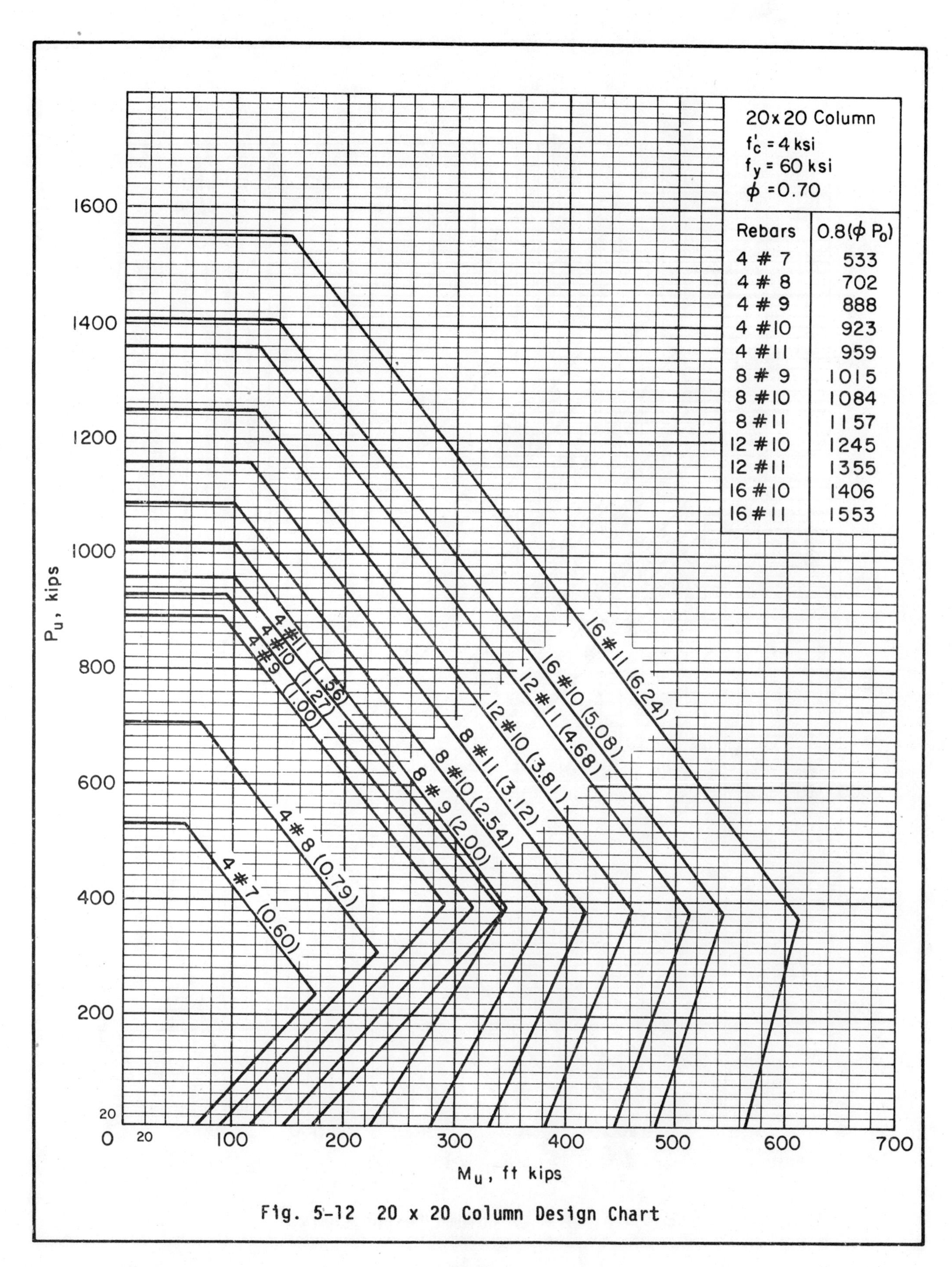

Fig. 5-12 20 x 20 Column Design Chart

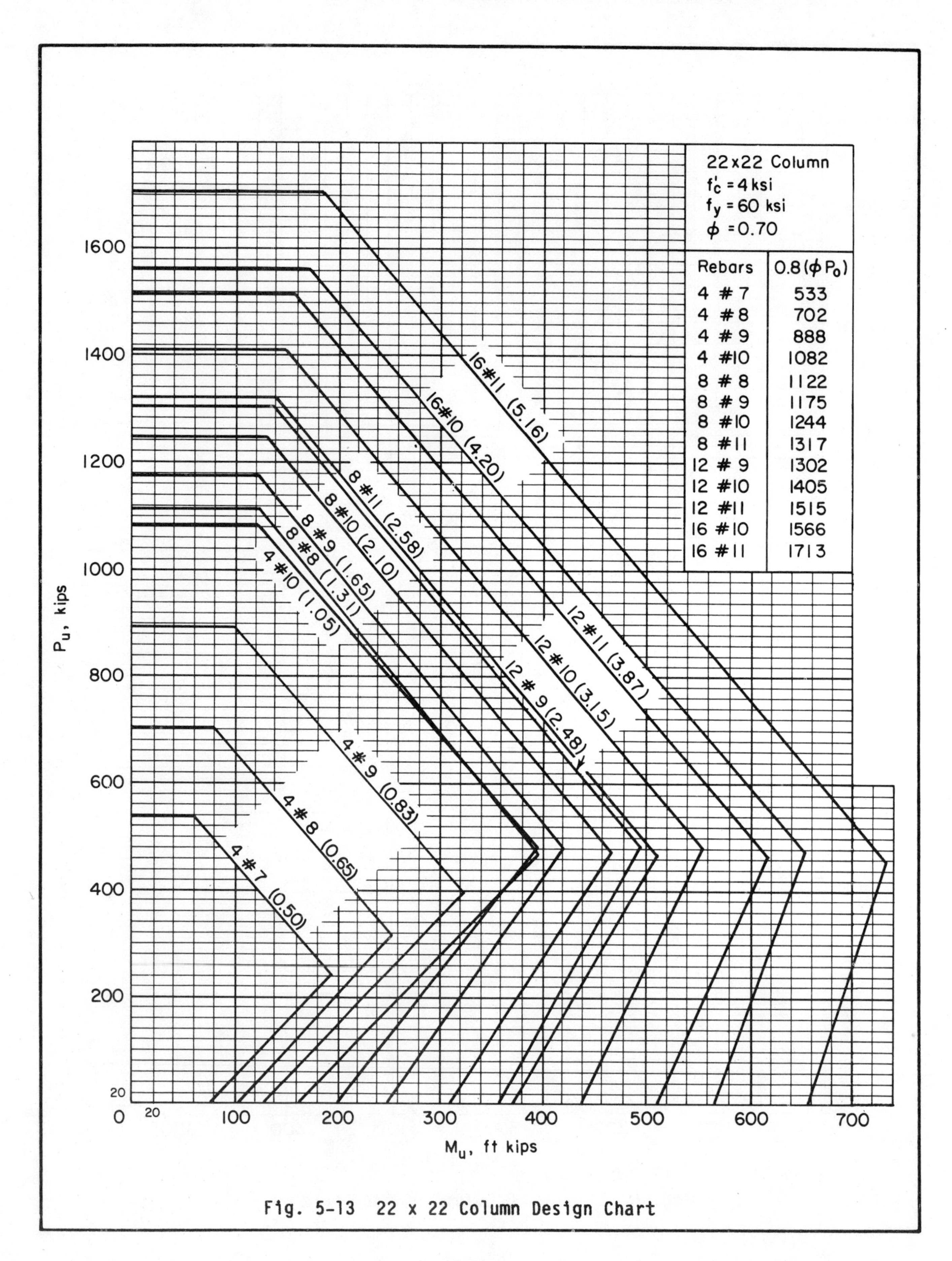

Fig. 5-13 22 x 22 Column Design Chart

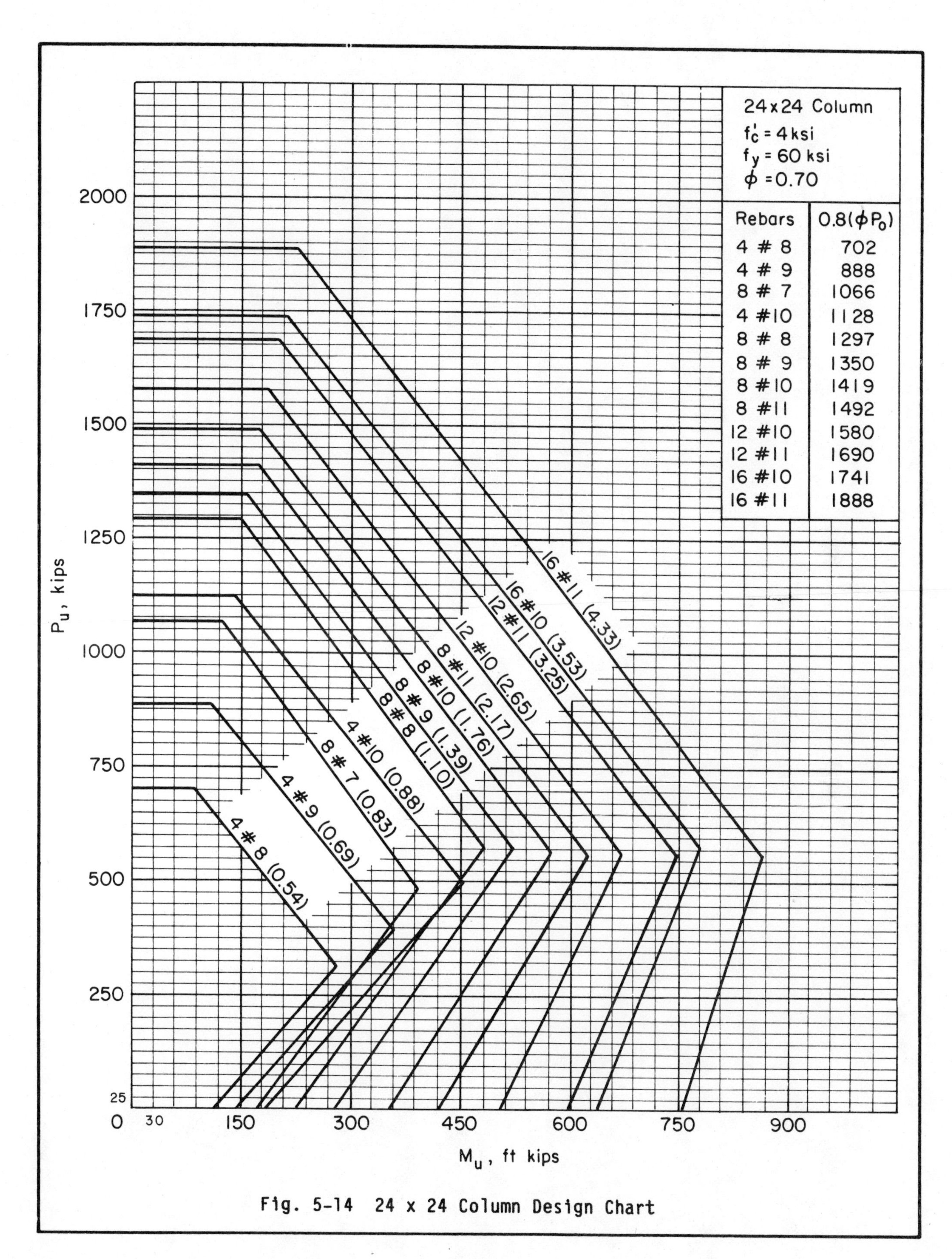

Fig. 5-14 24 x 24 Column Design Chart

6

Simplified Design for Structural Walls

Arnold H. Bock*

6.1 INTRODUCTION

For low-to-moderate height buildings, frame action alone usually suffices to provide lateral resistance. The 0.75 factor permitted for the load combination including wind effects [ACI Eq. (9-2)] is, in most cases, sufficient to accommodate the wind forces and moments without an increase in member sizes (columns) required to accommodate the lateral load effects of wind. Whether directly considered or not, nonstructural masonry walls and partitions can also add significantly to the total rigidity of a building and provide reserve capacity against lateral loads.

Use of structural walls to resist lateral load effects are certainly more pronounced in high-rise buildings to supplement frames that, if unaided, often could not be efficiently designed to satisfy lateral load requirements nor acceptable lateral drift requirements. Frame buildings depend primarily on frame action and the rigidity of member connections (slab-column or beam-column) for their resistance to lateral load effects and they tend to be uneconomical beyond 10 stories. To improve economy, structural walls are usually necessary in buildings exceeding 10 stories in height.

If structural walls are to be incorporated into the framing system, a tentative decision needs to be made at the conceptual design stage concerning

*Senior Regional Structural Engineer, South Central Region, PCA

location and plan layout. Many buildings are constructed with a central core area for vertical transportation (elevators and stairways) and utility services or a stairwell at one or two locations away from the core. All such service areas are enclosed with walls, and in some building occupancies the walls must be fire separation walls.

If concrete (or masonry) enclosure walls are used, such walls can be bifunctional and serve both as structural walls for lateral load resistance and bracing for the frame. In the conceptual stage, an effort should be made to locate the structural walls symmetrically within the plan layout of the building so that the center of lateral resistance of the walls coincides with the center of wind shear to avoid torsional effects. Concrete floor systems are sufficiently rigid to act as horizontal diaphragms to distribute the lateral loads in proportion to the stiffness of the vertical framing elements. The stiffening effect of structural walls also greatly improves the total rigidity of the building against lateral drift, especially when used with a flat-plate floor system without the stiffening effect of beams between column supports.

6.2 FRAME-WALL INTERACTION

The analysis and design of a building frame can be simplified if structural walls are incorporated into the structural framing system to provide both lateral load resistance and lateral bracing for the frame. For a dual framing system, the analysis is considerably simplified by assigning the total lateral load effects of wind to the walls, with the walls analyzed as simple cantilever spans. Members of the frame (columns and beams or slabs) are then proportioned to resist the vertical load effects of dead and live load only. Neglecting frame-wall load interaction for buildings of moderate size and height gives reasonable results both from an economic and structural consideration. With walls six times as stiff as the total columns within a story, the frame takes only a small portion of the lateral loads. Thus, for low-rise buildings, neglecting the contribution of frame action in resisting lateral loads and assigning the total lateral load resistance to walls is an entirely reasonable assumption. In contrast, frame-wall interaction must be considered for tall, tower-like structures where the walls have a pronounced structural effect on the frame, especially in the upper stories. Due to the interactive forces between the frame and walls, the frame is usually subjected to forces higher than the externally applied wind forces in the upper stories. Economics of tall structures also dictate an analysis as a dual lateral load resisting system.

With adequate wall bracing, the frame can be considered braced for column design with column slenderness effects usually neglected, except for very slender columns. Consideration of slenderness effects for braced and unbraced columns is discussed in Chapter 5, Section 5.5.

6.3 WALL SIZING FOR LATERAL BRACING

Practical considerations of size of opening required for stairwells and elevators will usually dictate a minimum wall plan layout. Also, as a practical matter in placing concrete for reinforced wall construction, a minimum wall thickness of 8 in. for a single layer of reinforcement and 10 in. minimum for a double layer will be required. Minimum wall thickness for required fire resistance of "enclosure" walls also must not be overlooked. See Chapter 10 for design considerations for fire resistance. The above considerations will, in most cases, provide a total wall area and stiffness sufficient for both lateral load resistance and lateral bracing to qualify the frame as braced.

A simple criteria is given in ACI Commentary 10.11.2 to establish whether structural walls provide sufficient lateral bracing to qualify the frame as braced: "Structural walls must have a total stiffness at least equal to six (6) times the sum of the stiffnesses of all the columns."

$$I_{(wall)} \geq 6\ [\Sigma I_{(int\ col)} + \Sigma I_{(ext\ col)}]$$

The above criteria is simple enough for sizing the structural walls so that the frame can be considered "braced."

6.3.1 Example: Wall Sizing for Braced Condition

Using the approximate six times stiffness criteria, size the structural walls for Alternate (2) of Building #2 (5-story flat plate). Note: The 5-story flat plate frame of Building #2 is certainly within the lower height range for structural wall consideration. The structural wall alternative is presented to illustrate the difference in analysis and design of a framing system with or without wall bracing. Both architectural and economic considerations need to be evaluated to effectively conclude for or against the use of structural walls in low-to-moderate height buildings on a case by case basis.

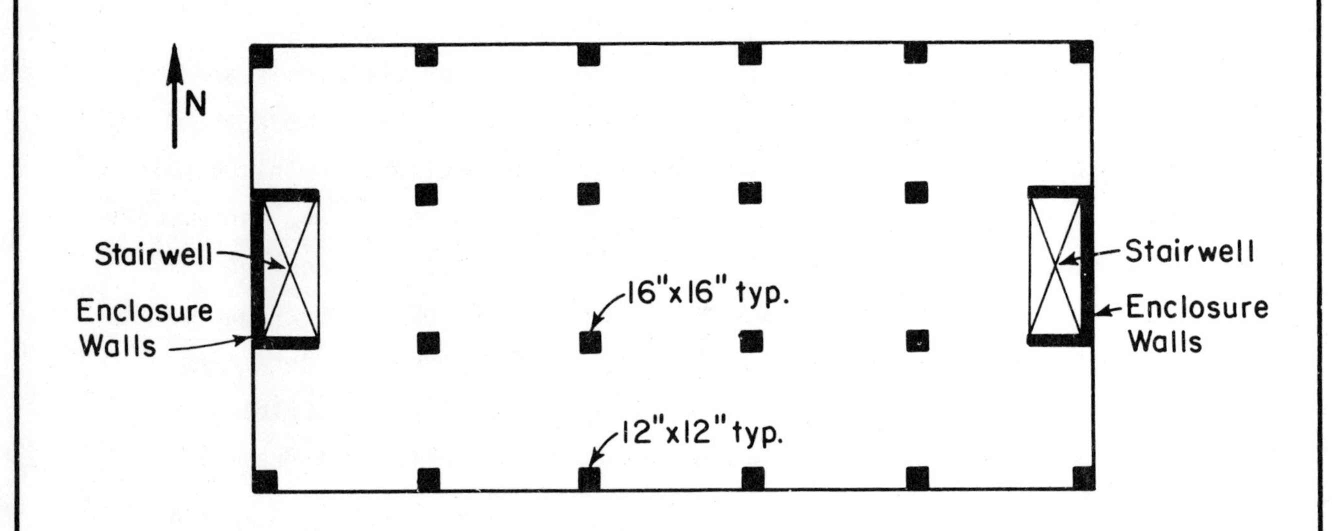

$$I_{(wall)} \geq 6\ [\Sigma I_{(int\ col)} + \Sigma I_{(ext\ col)}]$$

$$\Sigma I_{(int\ col)} = 8\ (16^4/12) = 43{,}691$$

$$\Sigma I_{(ext\ col)} = 12\ (12^4/12) = \underline{20{,}736}$$

$$64{,}427 \text{ in.}^4$$

$$I_{(wall)} \geq 6\ [\Sigma I_{(col)}] = 386{,}562 \text{ in.}^4$$

Use wall thickness = 8 in. (single layer of reinforcement). To accommodate opening required for stairwells, provide 8'-0 "flanges."

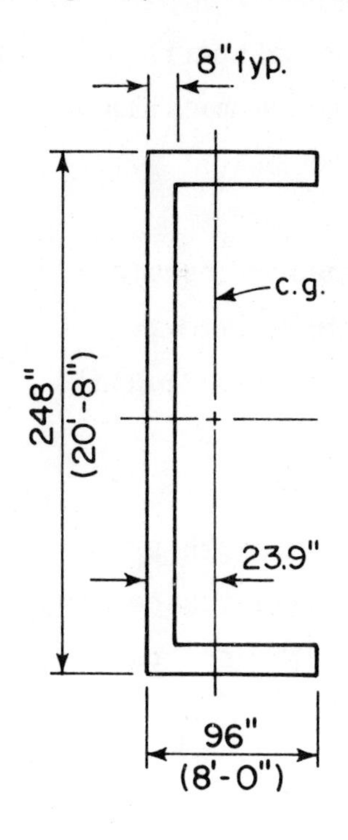

EW-direction

A_g = 248x8 + 88x8x2 = 1984 + 1408 = 3392 $in.^2$

$\bar{y}$ = (1984x4 + 1408x52)/3392 = 23.9 in.

I_g = [$248x8^3/12$ + $1984x19.9^2$] + [$2(8x88^3/12)$ + 1408×28.1^2] = 2,816,665 $in.^4$

For two end wall units:

I_g = 2 (2816665) = 5,663,330 » 386,562.

Therefore, the frame can be considered braced for column design. With the wall segments in the EW-direction providing most of the required stiffness in the EW-direction, the 8'-0 length required for stairwell enclosure is more than adequate.

6.4 DESIGN FOR WALL SHEAR STRENGTH

Design for horizontal shear forces (in the plane of the wall) is primarily of importance for structural walls with a small height-to-length ratio (walls in low-rise buildings). Design of higher walls, particularly walls with uniformly distributed reinforcement, will probably be governed by moment considerations. It is therefore essential that the moment strength of walls be computed, along with their shear strength. Flexural strength is discussed in Section 6.5.

A wall with near minimum amount of both vertical and horizontal reinforcement (uniformly distributed) is usually the most economical. If much more than the minimum amount of wall reinforcement is required to resist the design shear forces (and moments), a change in wall size (length or thickness) should be considered. Shear reinforcement for walls is covered by ACI 11.10. Both vertical and horizontal reinforcement must be provided in all walls. Amount of reinforcement (vertical and horizontal) is based on the magnitude of the design shear force:

(1) When the design shear force is less than 1/2 the shear strength provided by concrete ($V_u < \varphi V_c/2$), minimum wall reinforcement according to ACI 14.3 applies. Suggested vertical and horizontal reinforcement for this condition is given in Table 6-1.

(2) When the design shear force is less than the shear strength provided by concrete ($V_u < \varphi V_c$), minimum shear reinforcement according to ACI 11.10.9 applies. Suggested reinforcement (both vertical and horizontal) for this condition is given in Table 6-2.

(3) When the design shear force exceeds the concrete shear strength ($V_u > \varphi V_c$), horizontal shear reinforcement must be calculated with the minimum (both vertical and horizontal) not less than that given in Table 6-2.

Using the same basic approach as used in Section 3.6 for beams, design for required horizontal shear reinforcement can be simplified by assigning specific values to the two resisting shear force components ($\varphi V_c + \varphi V_s$) for our selected materials f'_c = 4000 psi and f_y = 60,000 psi, and a specific cross section. A specific value of φV_s can be derived for each bar size and spacing as follows:

$$\varphi V_s = \varphi \frac{A_v f_y d}{s_2} \qquad \text{ACI Eq. (11-34)}$$

Substituting φ = 0.85, f_y = 60,000 psi, and d = 0.8 ℓ_w (ACI 11.10.4); and for, say, #3 @ 6 in. (single layer):

$$\varphi V_s = 0.85(0.11)60\text{x}0.8(12\ell_w)/6 = 8.98\ell_w \text{ kips, say } 9.0\ell_w \text{kips}$$

where ℓ_w is horizontal length of wall resisting shear (in feet).

For a given length of wall ℓ_w , Table 6-3 gives values of φV_s per foot of wall length (τ_s) for selection of horizontal wall reinforcement with Grade 60 rebars.

In a similar manner, Table 6-4 gives design values for $\varphi V_c = \varphi 2 \sqrt{f'_c} h(0.8\ell_w)$ and limiting values for $\varphi V_n = \varphi V_c + \varphi V_s = \varphi 10 \sqrt{f'_c} h(0.8\ell_w)$, both expressed per foot of wall length (τ_c and τ_n).

Required amount of vertical shear reinforcement is simply a proportion of the horizontal shear reinforcement as given by ACI Eq. (11-35):

$$\rho_n = 0.0025 + 0.5\ (2.5 - h_w/\ell_w)(\rho_h - 0.0025)$$

where $\rho_n = A_v/s_1h$ = ratio of vertical shear reinforcement
$\rho_h = A_v/s_2h$ = ratio of horizontal shear reinforcement

When the wall height-to-~~thickness~~ length ratio h_w/ℓ_w (total height of wall from base to top/horizontal length of wall) is less than 0.5, the vertical reinforcement is made the same as the horizontal reinforcement. When h_w/ℓ_w is greater than 2.5, only the minimum amount of vertical reinforcement is required per Table 6-2. For intermediate values of h_w/ℓ_w, required ρ_n is calculated from the above equation.

6.4.1 Example: Design for Shear Strength

To illustrate simplified design for wall shear reinforcement, select wall reinforcement for a 10'-0 x 8 in. wall segment to resist a story shear of 200 kips. Assume total height of wall from base to top = 20'-0.

(1) Determine factored shear force. Use ACI Eq. (9-3) for wind only.

$$V_u = 1.3(200) = 260^k$$

Table 6-1 **Minimum Wall Reinforcement** ($V_u < \varphi V_c/2$)

Wall Thickness h (in.)	Horizontal Reinforcement*					
	Single Layer			Double Layer		
6	#3@9	#4@17				
8	#3@7	#4@12	#5@18	#3@13		
10		#4@10	#5@15	#3@11	#4@18	
12		#4@8	#5@13	#3@9	#4@16	
14		#4@7	#5@11	#3@8	#4@14	
16		#4@6	#5@9	#3@7	#4@12	#5@18
18			#5@8	#3@6	#4@11	#5@17
Wall Thickness h (in.)	**Vertical Reinforcement ****					
	Single Layer			Double Layer		
6	#3@15					
8	#3@11	#4@18				
10	#3@9	#4@16		#3@18		
12	#3@8	#4@13		#3@15		
14		#4@12	#5@18	#3@13		
16		#4@10	#5@16	#3@11		
18		#4@9	#5@14	#3@10	#4@18	

$*A_s = 0.002 s_2 h$ (ACI 14.3.3)
$**A_s = 0.0012 s_1 h$ (ACI 14.3.2)

Table 6-2 **Minimum Shear Reinforcement for Walls** ($V_u < \varphi V_c$)

Wall Thickness h (in.)	Horizontal & Vertical Reinforcement*					
	Single Layer			Double Layer		
6	#4@13					
8	#4@10	#5@15		#3@11	#4@18	
10	#4@8	#5@12	#6@17	#3@9	#4@16	
12	#4@6	#5@10	#6@14	#3@7	#4@13	
14		#5@8	#6@12		#4@11	#5@18
16		#5@7	#6@11		#4@10	#5@15
18		#5@6	#6@9			#5@14

$*A_s = 0.0025 sh$ (ACI 11.10.9)
Note: Also gives maximum spacing for a given wall thickness and bar size to provide $\rho_{min} = 0.0025$

Table 6-3 Shear Strength φV_s Provided by Horizontal Shear Reinforcement

$\tau_s = \frac{\phi V_s}{\ell_w}$ (kips/ft)

Bar Spacing s_2 (in.)	~~$\varphi V_s = \tau_s \ell_w$* (kips)~~			
	#3	#4	#5	#6
6	9.0	16.3	25.3	35.9
7	7.7	14.0	21.7	30.8
8	6.7	12.2	19.0	26.9
9	6.0	10.9	16.9	23.9
10	5.4	9.8	15.2	21.5
11	4.9	8.9	13.8	19.6
12	4.5	8.2	12.7	18.0
13	4.1	7.5	11.7	16.6
14	3.9	7.0	10.8	15.4
15	3.6	6.5	10.1	14.4
16	3.4	6.1	9.5	13.5
17	3.2	5.8	8.9	12.7
18	3.0	5.4	8.4	12.0

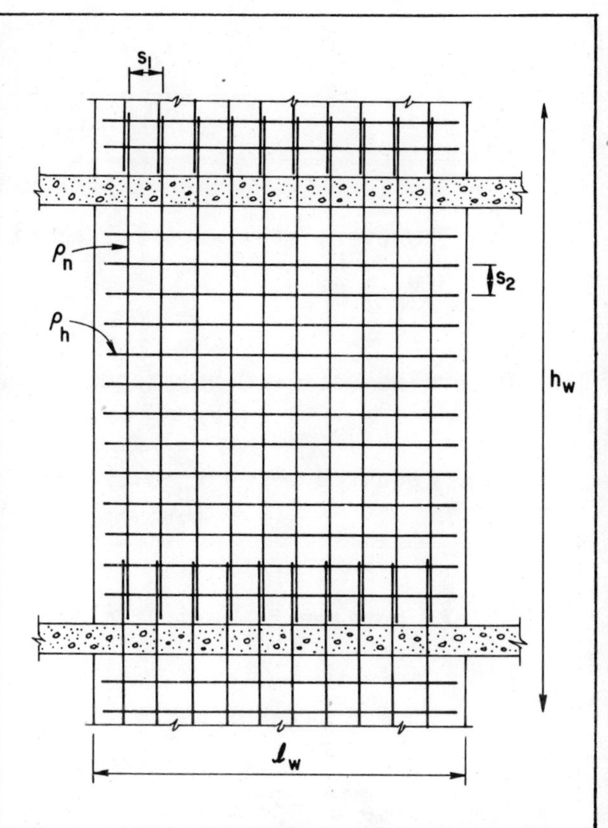

*Values are for single layer, (2x) for double layer. ℓ_w in feet, Grade 60 rebars.

Table 6-4 Design Values for φV_c and φV_n ~~(kips)~~ kips/ft*

Wall Thickness h (in.)	$\tau_c = \frac{\phi V_c}{\ell_w}$ ~~$\varphi V_c = \tau_c \ell_w$~~	$\tau_n = \frac{\phi V_c + \phi V_s}{\ell_w}$ ~~$\varphi V_c + \varphi V_s = \tau_n \ell_w$~~
6	6.2	31.0
8	8.3	41.3
10	10.3	51.6
12	12.4	61.9
14	14.5	72.3
16	16.5	82.6
18	18.6	92.9

*f'_c = 4000 psi ℓ_w in feet

(2) Determine shear strength φV_c and φV_n.

From Table 6-4: $\varphi V_c = 10(8.3) = 83^k$

$\varphi V_n = 10\ (41.3) = 413^k$

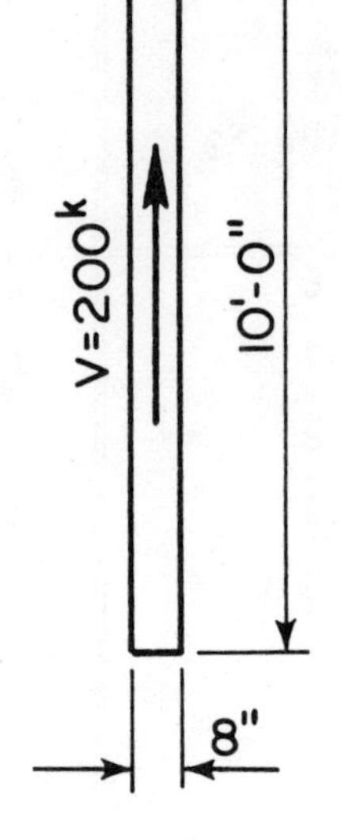

Wall cross section is adequate ($V_u < \varphi V_n$); however, shear reinforcement must be calculated ($V_u > \varphi V_c$).

(3) Determine required horizontal shear reinforcement.

required $\varphi V_s = V_u - \varphi V_c = 260 - 83 = 177^k$

$\varphi V_s = 177/10 = 17.7^k$/ft of wall length

Select horizontal bars from Table 6-3:

#5 @ 8 provide $\varphi V_s = 19.0^{k/'} > 17.7^{k/1}$ required.

Note: Use of minimum shear reinforcement for an 8 in. wall thickness is not adequate:
#5 @ 15 (Table 6-2) provide only $\varphi V_s = 10.1^{k/'}$ (Table 6-3).

(4) Determine required vertical shear reinforcement.

$\rho_n = 0.0025 + 0.5(2.5 - h_w/\ell_w)(\rho_h - 0.0025)$ ACI Eq.(11-35)

$= 0.0025 + 0.5(2.5 - 2)(0.0048 - 0.0025)$

$= 0.0031$

where $h_w/\ell_w = 20/10 = 2$

$\rho_h = A_v/sh = 0.31/8 \times 8 = 0.0048$

required $A_v/s = \rho_n h = 0.0031 \times 8 = 0.0248$

For #5 bars: $s_{max} = 0.31/0.0248 = 12.5$ in.

Bar Selection for Shear Strength

Vertical Bars: #5 @ 12

Horizontal Bars: #5 @ 8

Note: Table 6-4 can also be used directly to determine required length of wall segment to resist a given shear force using minimum shear reinforcement ($V_u < \varphi V_c$). For our example, length of wall required to provide minimum shear reinforcement 260/8.3 = 31.3 ft.

6.4.2 Example: Design for Shear Strength

For Alternate (2) of Building #2 (5-story flat plate), select shear reinforcement for the two end shear wall units. Assume total wind forces are resisted by the structural walls with slab-column framing to resist gravity loads only.

(1) EW-direction

Total shear force at base of building (1st story); see Chapter 2, Section 2.2.1.1: (p 2-5)

$V = 11.3 + 22.0 + 21.1 + 19.9 + 20.5 = 94.8^k$

$V = 94.8/2 = 47.4^k$ (each end wall unit)

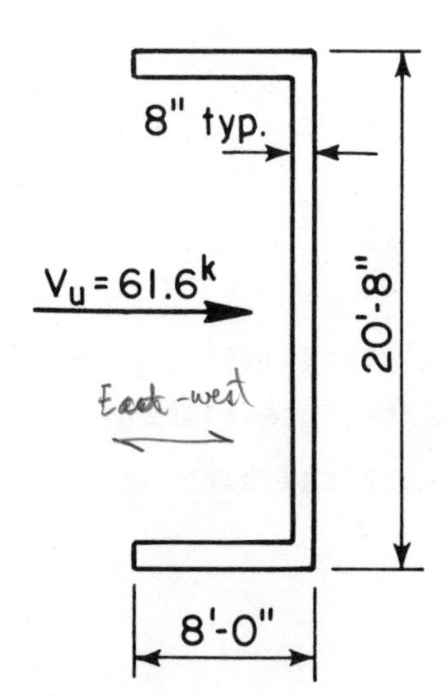

Factored shear force (Use ACI Eq. (9-3)for wind load only):

$$V_u = 1.3(47.4) = 61.6^k$$

For EW-direction, assume shear force resisted by two 8'-0 wall segments. For each ℓ_w = 8'-0 wall segment, shear strength φV_c (Table 6-4): (p 6-9)

$$\varphi V_c = 8(8.3) = 66.4^k$$

$$\frac{V_u}{\varphi V_c} = \frac{30.8}{66.4} = 0.46$$

Condition (1) applies ($V_u < \varphi V_c/2$); provide minimum wall reinforcement. Select reinforcement from Table 6-1. For 8 in. wall thickness, use either #3 @ 7, #4 @ 12, or #5 @ 18 for horizontal bars and either #3 @ 11 or #4 @ 18 for vertical bars in 8'-0 wall segments.

(2) NS-direction

Total shear force at base of building (1st story); see Chapter 2, Section 2.2.1.1:

$$V = 22.6 + 43.9 + 42.2 + 39.8 + 41.0 = 189.5^k$$
$$V = 189.5/2 = 94.8^k \text{ (each end wall unit)}$$

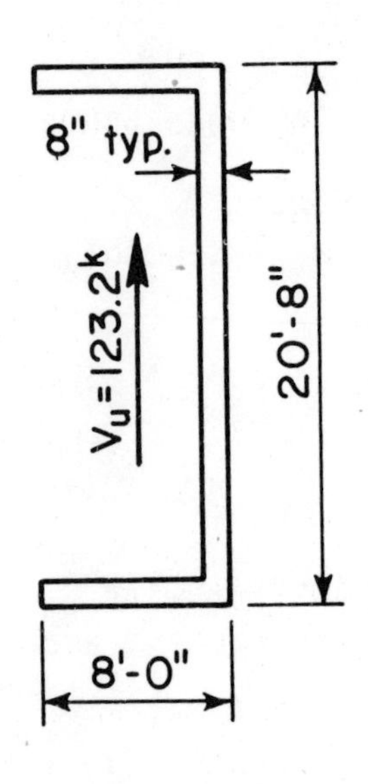

Factored shear force [ACI Eq. (9-3)]:

$$V_u = 1.3(94.8) = 123.2^k$$

For NS-direction, assume shear force resisted by 20'-8" wall segment. For $\ell_w = 20.67$, shear strength φV_c (Table 6-4):

$$\varphi V_c = 20.67(8.3) = 171.6^k$$

$$\frac{V_u}{\varphi V_c} = \frac{123.2}{171.6} = 0.72$$

Condition (2) applies ($V_u < \varphi V_c$); provide minimum shear reinforcement. Select reinforcement from Table 6-2. For 8 in. wall thickness, use either #4 @ 10 or #5 @ 15 (single layer) for both horizontal and vertical bars in 20'-8" segments.

(3) Check shear conditions for 2nd story in NS-direction.

$$V_u = 1.3(22.6 + 43.9 + 42.2 + 39.8)/2 = 96.5^k$$

$$\frac{V_u}{\varphi V_c} = \frac{96.5}{171.6} = 0.56 > 0.50$$

Minimum shear reinforcement is still required for the 20'-8" wall segments within the second story. For the 3rd story and above, minimum wall reinforcement as given in Table 6-1 could be used for all wall segments. For horizontal bars: #3 @ 7, #4 @ 12, or #5 @ 18 are possible choices; and for vertical bars: #3 @ 11 or #4 @ 18.

Bar Selection for Shear Strength

Vertical bars: Use #4 @ 9* for 1st and 2nd stories
#4 @ 18 for 3rd through 5th

Horizontal bars: Use #4 @ 10 for 1st and 2nd
#4 @ 12 for 3rd through 5th

*Note: For moment strength, #6 @ 9 are required for the 8'-0 wall segments within the first story. See Example 6.5.1.

For above example Final rebar arrangement see p6-20

6.5 DESIGN FOR WALL MOMENT STRENGTH

For structural walls in moderate height buildings, walls of uniform cross section with uniformly distributed vertical and horizontal reinforcement are usually the most economical. Concentration of reinforcement at the extreme ends of a wall (or wall segment) is usually not required for walls in moderate height buildings. Uniform distribution of the vertical wall reinforcement, as required for shear, will usually suffice for required moment strength. Also, minimum amount of reinforcement will usually be sufficient, not only for shear strength, but also for moment strength. Moment strength of a rectangular wall section containing uniformly distributed vertical reinforcement and subjected to combined moment and axial load can be easily calculated by:[6.1]

$$\varphi M_n = \varphi[0.5A_{st}f_y\ell_w(1 + \frac{P_u}{A_{st}f_y})(1 - \frac{c}{\ell_w})]$$

where A_{st} = total area of vertical wall reinforcement

$= A_b\ell_w/s$

ℓ_w = horizontal length of wall

s = spacing of vertical wall reinforcement

P_u = factored axial compressive load

$\frac{c}{\ell_w} = \frac{\omega + \alpha}{2\omega + 0.85\beta_1}$, where $\beta_1 = 0.85$ for $f'_c = 4000$

$\omega = (\frac{A_{st}}{\ell_w h})\frac{f_y}{f'_c}$

$\alpha = \frac{P_u}{\ell_w h f'_c}$

h = overall thickness of wall

φ = 0.90 (strength primarily controlled by flexure with low axial load.)

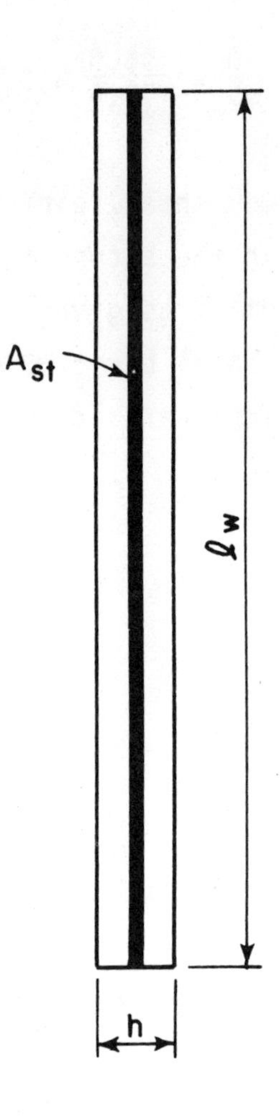

6.5.1 Example: Design for Wall Moment Strength

For Alternate (2) of Building #2 (5-story flat plate), design moment reinforcement for the two end shear wall units. Assume 8'-0 wall segments resist wind moments in EW-direction and 20'-8 wall segments resist wind moments in NS-direction. Roof: DL = 122 psf Floors: DL = 142 psf

(1) Factored loads and load combinations to be considered.

For evaluating wall moment strength, wind load combination ACI Eq. (9-3) will govern.

$U = 0.9D + 1.3\ W$

(a) Dead load for each end shear wall unit at first story level.

Tributary floor area = 12x20x2 = 480 sq ft/story

Wall dead load = 3392x0.150/144 = 3.53^k/ft height

$P_u = 0.9(0.122 \times 480 + 0.142 \times 480 \times 4 + 3.53 \times 63) = 498^k$

Proportion total P_u between wall segments:

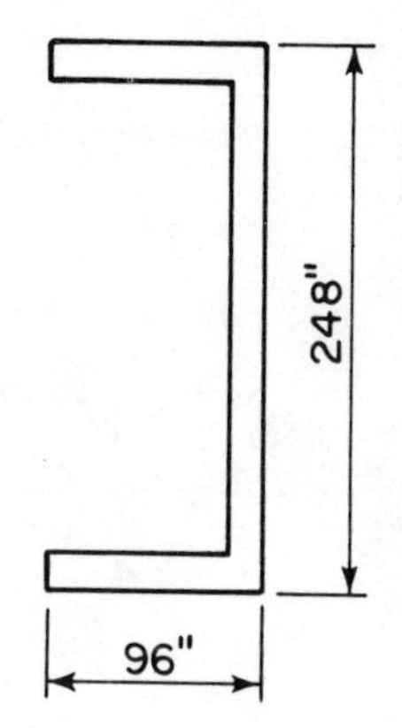

2 - 8'-0 Segments = 2x96 = 192		192/440 = 0.44
1 - 20'-8 Segment	= 248	248/440 = 0.56

for 2 - 8'-0 Segments, $P_u = 0.44(498) = 219^k$

1 - 20'-8 Segment, $P_u = 0.56(498) = 279^k$

(b) Wind moments for each end shear wall unit at first story level.

From wind load analysis; see Chapter 2, Section 2.2.1.1:

EW-direction

$$M_u = 1.3(11.3x63 + 22x51 + 21.1x39 + 19.9x27 + 20.5x15)/2$$
$$= 2276'^k$$

NS-direction

$$M_u = 1.3(22.6x63 + 43.9x51 + 42.2x39 + 39.8x27 + 41x15)/2$$
$$= 4549'^k$$

(c) Values of P_u and M_u for the 2nd and 3rd story levels are obtained in a similar manner.

For 2nd story level: 2 - 8'-0 Segments $P_u = 171^k$

1 - 20'-8 Segment $P_u = 218^k$

$$EW - M_u = 1352^{'k}$$
$$NS - M_u = 2701^{'k}$$

For 3rd story level: 2 - 8'-0 Segments $P_u = 128^k$
1 - 20'-8 Segment $P_u = 162^k$

$$EW - M_u = 772^{'k}$$
$$NS - M_u = 1543^{'k}$$

(2) Design for wall moment strength in EW-direction

Determine moment strength for wall segments in EW-direction. Initially check moment strength for required vertical shear reinforcement #4 @ 9 (see Example 6.4.2).

(a) For 2 - 8'-0 wall segments at first story level.

$P_u = 219^k$

$M_u = 2276^{'k}$

$\ell_w = 96"$

$h_{eff} = 2(8) = 16"$

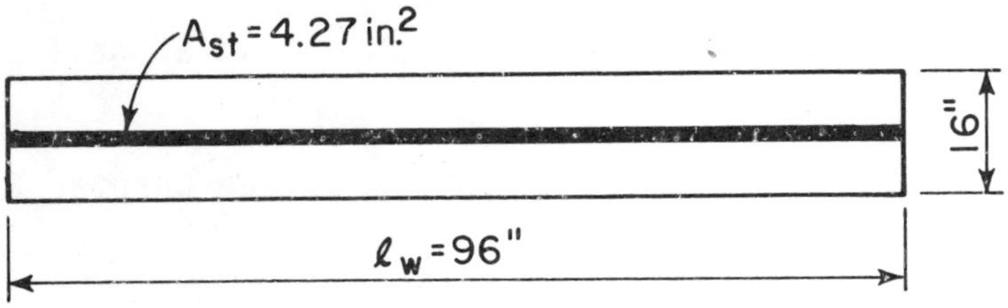

$A_{st} = 4.27$ in.^{2}

16"

$\ell_w = 96"$

For #4 @ 9 (2 wall segments):

$$A_{st} = A_b \ell_w / s = 0.20x2x96/9 = 4.27 \text{ in.}^2$$

$$\omega = \left(\frac{A_{st}}{\ell_w h}\right) \frac{f_y}{f'_c} = \left(\frac{4.27}{96x16}\right) \frac{60}{4} = 0.042$$

$$\alpha = \frac{P_u}{\ell_w h f'_c} \quad \frac{219}{96x16x4} = 0.036$$

$$\frac{c}{\ell_w} \quad \frac{\omega + \alpha}{2\omega + (0.85)^2} = \frac{0.042 + 0.036}{2(0.042) + 0.72} = 0.097$$

$$M_n = 0.5 A_{st} f_y \ell_w \left(1 + \frac{P_u}{A_{st} f_y}\right)\left(1 - \frac{c}{\ell_w}\right)$$

$$= 0.5x4.27x60x96(1+\frac{219}{4.27x60})(1-0.097)/12 = 1716^{'k}$$

$$\varphi M_n = 0.9(1716) = 1544^{'k} < 2276 \quad \text{N.G.}$$

The #4 @ 9 are not adequate for wind moment strength in the EW-direction at the first story level.

Moment strength for #5 @ 9 is also less than the required M_u ($\varphi M_n = 1952^{'k} < 2276$); calculations not shown here.

Try #6 @ 9: $A_{st} = 0.44 \times 2 \times 96/9 = 9.39 \text{ in.}^2$

$$\omega = (\frac{9.39}{96x16}) \frac{60}{4} = 0.092$$

$$\frac{c}{\ell_w} = \frac{0.092 + 0.036}{2(0.092)+0.72} = 0.142$$

$$M_n = 0.5x9.39x60x96(1+\frac{219}{9.39x60})(1-0.142)/12 = 2685^{'k}$$

$$\varphi M_n = 0.9(2685) = 2417^{'k} > 2276 \quad \text{OK}$$

(b) For 2 - 8'-0 wall segments at 2nd story level.

$P_u = 171^k$
$M_u = 1352^{'k}$

Check #4 @ 9: $A_{st} = 4.27 \text{ in.}^2$

$$\omega = 0.042$$

$$\alpha = \frac{171}{96x16x4} = 0.028$$

$$\frac{c}{\ell_w} = \frac{0.042 + 0.028}{2(0.042)+0.72} = 0.087$$

$$M_n = 0.5x4.27x60x96(1+\frac{171}{4.27x60})(1-0.087)/12 = 1560^{'k}$$

$$\varphi M_n = 0.09(1560) = 1404^{'k} > 1352 \quad \text{OK}$$

The #4 @ 9 (required shear reinforcement) are adequate for wind moments above the first story.

(c) For two 8'-0 wall segments at 3rd story level.

$P_u = 128^k$
$M_u = 772^{'k}$

Check #4 @ 18 (required shear reinforcement above 2nd story):

$$A_{st} = 0.20 \times 2 \times 96/18 = 2.13 \text{ in.}^2$$

$$\omega = \left(\frac{2.13}{96x16}\right) \frac{60}{4} = 0.021$$

$$\alpha = \left(\frac{128}{96x16x4}\right) = 0.021$$

$$\frac{c}{\ell_w} = \frac{0.021+0.021}{2(0.021)+0.72} = 0.055$$

$$M_n = 0.5x2.13x60x96\left(1+\frac{128}{2.13x60}\right)(1-0.055)/12 = 967^{'k}$$

$$\varphi M_n = 0.9(967) = 870^{'k} > 772 \quad \text{OK}$$

The #4 @ 18 (required shear reinforcement) are adequate for wind moments above the 2nd story.

(3) Design for wall moment strength in NS-direction

Determine moment strength for wall segments in NS-direction. Initially check moment strength for required vertical shear reinforcement #4 @ 9 (see Example 6.4.2)

(a) For 1 - 20'-8 wall segment at first story level.

$P_u = 279^k$
$M_u = 4549^{'k}$
$\ell_w = 248"$
$h = 8"$

For #4 @ 9: $A_{st} = 0.20x248/9 = 5.51$ in.^{2}

$$\omega = (\frac{5.51}{248x8})\ \frac{60}{4} = 0.042$$

$$\alpha = \frac{279}{248x8x4} = 0.035$$

$$\frac{c}{\ell_w} = \frac{0.042 + 0.035}{2(0.042)+0.72} = 0.096$$

A_{st}=5.51 in.^{2}

ℓ_w=248"

8"

$$M_n = 0.5x5.51x60x248(1+\frac{279}{5.51x60})(1-0.096)/12 = 5695^{'k}$$

$$\varphi M_n = 0.9(5695) = 5126^{'k} > 4549 \quad OK$$

(b) For 1 - 20'-8 wall segment at 3rd story level.

$P_u = 162^k$
$M_u = 1543^{'k}$

Check #4 @ 18 (required shear reinforcement above 2nd story).

$$A_{st} = 0.20x248/18 = 2.76 \text{ in.}^2$$

$$\omega = (\frac{2.76}{248x8})\ \frac{60}{4} = 0.021$$

$$\alpha = \frac{162}{248x8x4} = 0.020$$

$$\frac{c}{\ell_w} = \frac{0.021 + 0.020}{2(0.021)+0.72} = 0.054$$

$$M_n = 0.5x2.76x60x248(1+\frac{162}{2.76x60})(1-0.054)/12 = 3202^{'k}$$

$$\varphi M_n = 0.9(3202) = 2882^{'k} > 1543 \quad OK$$

The required shear reinforcement for the 20'-8 wall segments is adequate for wind moments for full height of building.

(4) Summary

Required shear reinforcement (Example 6.4.2) is adequate for wall moment strength except for the 8'-0 wall segments within the 1st story where #6 @ 9 are required.

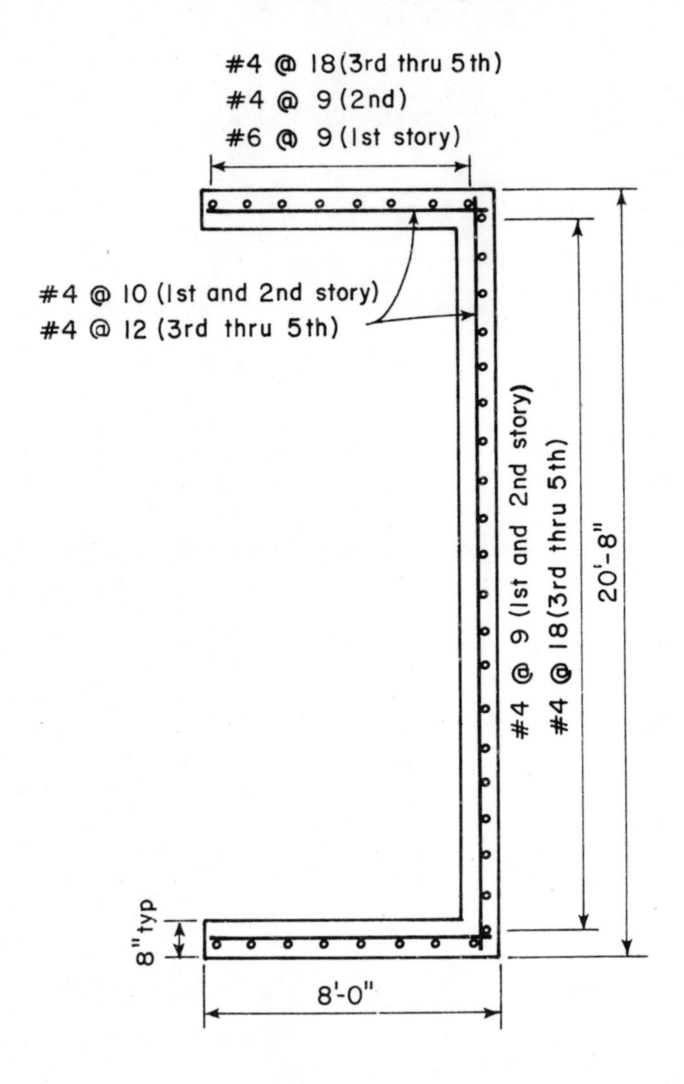

Selected Reference

6.1 Cardenas, A.E., Hanson, J.M., Corley, W.G., Hognestad, E., "Design Provisions for Shear Walls," Journal, American Concrete Institute, Proceedings Vol. 70, No. 3, March 1973, pages 221-230. Also RD Bulletin RD028.010, Portland Cement Association.

7

Simplified Design for Footings

Gary D. Pfuehler*

7.1 INTRODUCTION

A simplified method for design of spread footings is presented that will enable the designer to obtain required footing thickness with a one-step design equation based on minimum footing reinforcement. Also included is a rapid solution for bearing analysis, for footing dowels, and for horizontal load transfer at the base of a column. The methodology is confined to the use of individual square footings supporting square (or circular) columns and subject to uniform soil pressure.

The design method is intended to address the usual design conditions for footings of low-to-moderate height buildings. Design conditions such as uplift and overturning, if applicable, are beyond the scope of the simplified method and will need to be investigated further.

Also included is a simplified one-step thickness design equation for plain concrete footings. See discussion under Section 7.2.

Choice of Footing Materials....for footings, a concrete strength of f'_c = 3000 psi is the most common and economical choice. Higher strength concrete can be used where footing depth or weight is critical and must be minimized, but usually, savings in concrete volume do not offset higher unit price.

*Regional Structural Engineer, Great Lakes-Midwestern Region, PCA

Where concrete strength is a factor in the simplified design procedure, data are presented for both 3000 and 4000 psi concrete strengths. And, of course, Grade 60 bars are the standard grade recommended for overall economy.

7.2 PLAIN VS. REINFORCED FOOTINGS

Often reinforced footings are used in smaller buildings without considerations being given to the possibilities of a more economical alternative, plain footings. Many factors need to be considered when comparing the two alternatives, the most important being economic considerations. Among the other factors, such as soil type and job-site conditions and building size (loadings to be transferred), choice between plain and reinforced involves a "material trade-off" between concrete and reinforcement. The current market price of concrete and reinforcement are important decisionmaking parameters. "Time is money" is often heard in the construction industry. If plain footings can save considerable construction time, then the extra concrete may be economically justified. Also, savings in reinforcement and in placement may offset the additional concrete cost for a plain footing alternative.

For a given project, both plain and reinforced footing alternatives can be quickly proportioned by the simplified procedures and an overall cost comparison made, including both material and construction costs. For equivalent loading conditions, the thickness of a plain footing will be about twice that of a reinforced footing with minimum footing reinforcement. See Section 7.8.

7.3 SOIL PRESSURE

Soil pressures are usually recommended by a soil mechanics expert to the designer or set by local building code. In cities where experience and tests have established the allowable (safe) soil pressures of various foundation soils, local building codes may be consulted to determine the bearing capacities to be used in design. In the absence of such information or for conditions where the nature of the soils is unknown, borings or load tests should be made. For larger buildings, borings or load tests should always be made. Table 7-1 is given as a convenient reference. The bearing capacities given are average values compiled from a number of building codes.

Table 7-1 Average Bearing Capacities of Various Foundation Beds, in kips per sq ft (ksf)

Foundation bed	ksf
Alluvial soil	1
Soft clay	2
Firm clay	4
Wet sand	4
Sand and clay mixed	4
Fine dry sand	6
Hard clay	8
Coarse dry sand	8
Gravel	12
Gravel and sand, well cemented	16
Hard pan or hard shale	20
Medium rock	40
Hard rock	160

With allowable (safe) soil pressures usually applied to service load conditions, both service loads and factored loads become functions of the total footing design. The designer is required to select base area of footing using unfactored loads and allowable soil pressures, then perform the structural design (footing thickness and reinforcement) using factored loads. If the designer does not wish to accumulate both factored and unfactored loads for multistack columns, the allowable soil pressure q_a can be conservatively increased by a "load factor" of 1.6 (thus $q_s = 1.6\ q_a$) and use factored loads for the total design. Since determination of footing dimensions and footing reinforcement are generally not subject to precise analysis, use of the overall load factor is accurate enough for ordinary buildings.

7.4 SURCHARGE

In cases where the top of the footing is appreciably below grade or below the underside of a floor slab placed in contact with the ground, allowance needs to be made for the weight of earth replaced on top of the footing. In general, an allowance of 100 lb per cu ft is adequate for soil surcharge, unless wet packed conditions exist that warrant a higher value, say 120 lb per cu ft. Total surcharge (or overburden) above base of footing includes surface surcharge from floor slab (if any) plus soil surcharge (if any) plus footing weight.

7.5 ONE-STEP THICKNESS DESIGN FOR REINFORCED FOOTINGS * per 1 foot width *

A simplified footing thickness equation can be derived for individual footings with minimum reinforcement using the strength design data developed in Reference 7.1. For our selected footing materials, f'_c = 3000 psi and f_y = 60,000 psi, and with minimum reinforcement (ACI 10.5.3):

Set ρ = 0.0018 x 1.11* = 0.002

*correction for ratio of effective depth d to overall thickness h, assumed as $d/h \simeq 0.9$.

$$R_n = \rho f_y \left(1 - \frac{0.5\rho f_y}{0.85 f'_c}\right)$$

$$= 0.002 \times 60{,}000 \left(1 - \frac{0.5 \times 0.002 \times 60{,}000}{0.85 \times 3000}\right)$$

$$= 117.2 \text{ psi}$$

For design per foot width of footing:

$$d^2(\text{req'd}) = \frac{M_u}{\varphi R_n} = \frac{M_u \times 1000}{0.9 \times 117.2} = 9.48\, M_u \quad (M_u \text{ in ft-kips})$$

Referring to Fig. 7-1:

$$M_u = q_s \left(\frac{c^2}{2}\right) = \frac{P_u}{A_f}\left(\frac{c^2}{2}\right)$$, where c is footing projection from face of column (or wall).

$$d^2(\text{req'd}) = 4.74\, q_s c^2 = 4.74\, \frac{P_u c^2}{A_f}$$

$$d = 2.18\sqrt{q_s c^2} = 2.18\sqrt{\frac{P_u c^2}{A_f}}$$

With moment strength controlled by reinforcement, the above expression is equally applicable for other concrete strengths.

For $f_y = 60{,}000$:

$$d = 2.2\sqrt{q_s c^2} = 2.2\sqrt{\frac{P_u c^2}{A_f}}$$

Note: The above sizing equation is in mixed units; P_u in kips, A_f in square feet, and d in inches.

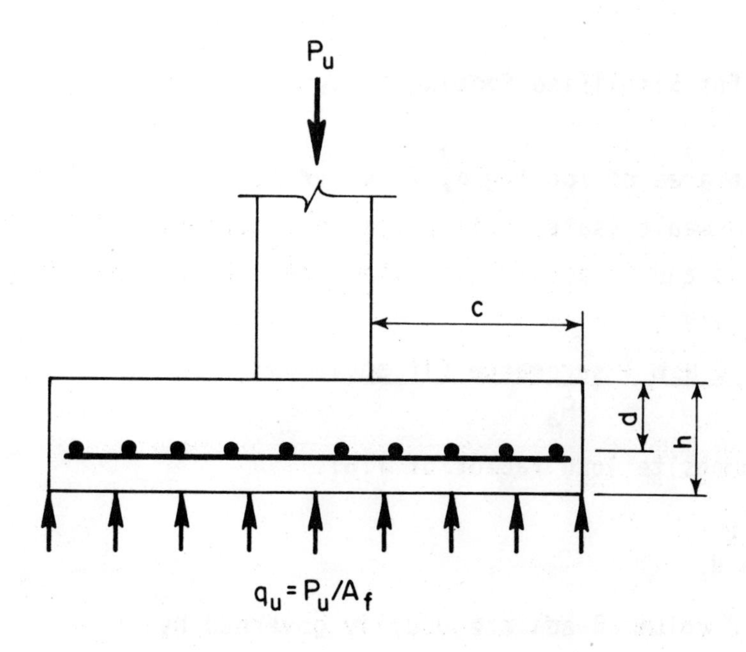

Fig. 7-1 Reinforced Footing

Uplift on "heel" ≡ negative

The one-step thickness equation for required moment strength of a footing with minimum reinforcement (ρ_{min} = 0.0018) applies for both square and rectangular footings (using largest value of c) and wall footings. With strength controlled primarily by reinforcement (f_y = 60,000 psi), the same equation is equally applicable for lesser or greater concrete strengths. Note that the simplified design assumes a uniform soil pressure distribution; for footings subject to axial load plus moment, an equivalent uniform soil pressure can be used.

Comment on Shear Strength -- The one-step thickness design equation, based on minimum reinforcement, results in a footing thickness that eliminates the need to consider shear strength for square footings supporting square (or circular) columns. For other footing and column shapes, shear strength may control footing thickness and will need to be investigated according to ACI 11.11. An experienced designer will soon recognize where shear is not critical and will adjust his design procedure accordingly. The simplified design is predicated on using square footings supporting square (or circular) columns.

7.5.1 - Procedure for Simplified Footing Design

(1) Determine base area of footing A_f from service loads (unfactored loads) and allowable (safe) soil pressure q_a determined for the site soil conditions and in accord with the local general building code.

$$A_f = \frac{D + L + W + \text{surcharge (if any)}}{q_a}$$

or using a composite load factor of 1.6:

$$A_f = \frac{P_u}{1.6\ q_a}$$

where factored column loads are usually governed by

$P_u = 1.4D + 1.7L$ ACI Eq. (9-1)

$P_u = 0.75(1.4D + 1.7L + 1.7W)$ ACI Eq. (9-2)

(2) Determine required footing thickness from one-step thickness equation.

$h = d + 4$ in. (3 in. cover + 1 bar dia. ≃ 1 in.) ← even w/ epoxy bars

$$h = 2.2\sqrt{\frac{P_u c^2}{A_f}} + 4 \text{ in.} \geq 6 \text{ in. (ACI 15.7)}$$

where P_u = factored column load, kips

A_f = base area of footing, sq ft

c = greatest distance from face of column to edge of footing, ft

h = overall thickness of footing, in.

Note: For square footings supporting square (or circular) columns, shear strength need not be checked. For other footing and column shapes, check depth required for shear strength according to ACI 11.11.

(3) Determine reinforcement.

$A_s = 0.0018\ bh$

For A_s per foot of footing width:

$A_s = 0.022h$ (in.^{2}/ft)

Select bar size and spacing from Table 7-2.

Table 7-2 - Areas or Bars per Foot Width of Footing - A_s (in.^{2}/ft)

BAR SIZE	BAR SPACING - inches												
	6	7	8	9	10	11	12	13	14	15	16	17	18*
# 3	0.22	0.19	0.17	0.15	0.13	0.12	0.11	0.10	0.09	0.09	0.08	0.08	0.07
# 4	0.39	0.34	0.29	0.26	0.24	0.22	0.20	0.18	0.17	0.16	0.15	0.14	0.13
# 5	0.61	0.53	0.46	0.41	0.37	0.34	0.31	0.29	0.27	0.25	0.23	0.22	0.21
# 6	0.88	0.76	0.66	0.59	0.53	0.48	0.44	0.41	0.38	0.35	0.33	0.31	0.29
# 7	1.20	1.03	0.90	0.80	0.72	0.65	0.60	0.55	0.51	0.48	0.45	0.42	0.40
# 8	1.57	1.35	1.18	1.05	0.94	0.85	0.78	0.72	0.67	0.62	0.59	0.55	0.52
# 9	2.00	1.71	1.50	1.33	1.20	1.09	1.00	0.92	0.86	0.80	0.75	0.71	0.67
#10	2.53	2.17	1.89	1.69	1.52	1.39	1.27	1.17	1.09	1.02	0.95	0.90	0.85
#11	3.12	2.68	2.34	2.08	1.87	1.70	1.56	1.44	1.34	1.25	1.17	1.10	1.04

*s_{max} = 18 in. (ACI 7.6.5)

Bar size must be small enough to be fully developed by anchorage between face of column and outer end of bar. Regardless of the development length required, the outer ends of the bars should not be less than 3 in. nor more than 6 in. from face of footing. Bar length ℓ_d is given in Table 7-3. If sufficient length is not available to develop a bar size selected, a smaller bar size (at closer spacing) should be selected to reduce the required ℓ_d. The following condition must be satisfied;

$$\ell_f \geq 2\ell_d + \text{column size} + 6\text{ in.}$$

Table 7-3 - Minimum Straight Tension Development Length ℓ_d (inches) for Grade 60 Bars

Bar Size	f'_c = 3000	f'_c = 4000
# 4	12	12
# 5	12	12
# 6	15	14
# 7	21	18
# 8	28	24
# 9	35	30
#10	45	38
#11	54	47

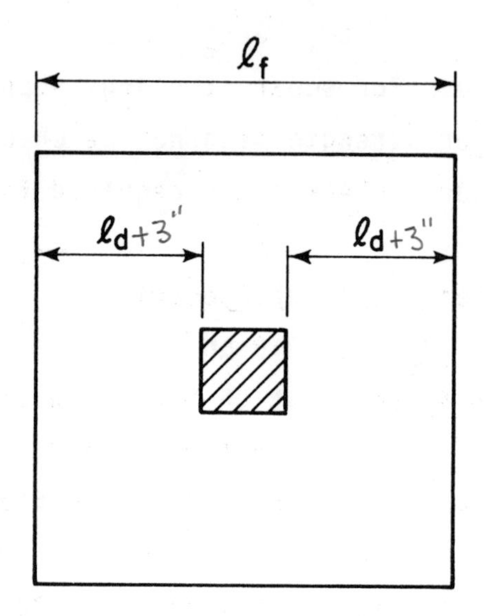

Available development length for footing reinforcement.

7.6 FOOTING DOWELS

7.6.1 - Vertical Force Transfer at Base of Column

The following design data addresses footing dowels designed to transfer calculated compression only. Dowels required to transfer calculated tension between column and footing must be anchored for tension with tension embedment and splice lengths. When footing dowels are designed for critical tension and spliced with column bars, splice length depends on class of splice required. In the usual case with all footing dowels spliced at one location, a Class C splice is required (ACI 12.15.1). If required loading conditions include uplift, the total tensile force must be transferred by dowels. Lateral forces must be transferred by shear-friction reinforcement or other means. See Section 7.6.2.

Calculated compression must be transferred by bearing on concrete and by dowels. Bearing strength must be adequate for both column concrete and footing concrete. Bearing on column concrete will always govern until the strength of column concrete exceeds twice that of the footing (ACI 10.15.1). For a column concrete strength of f'_c = 4000 psi, compression transferred by bearing on concrete is equal to φP_{nb} = 2.38 A_g, where A_g is gross area of column (ACI 10.15.1). For our selected column sizes, φP_{nb} is tabulated in Table 7-4.

Table 7-4 Bearing Capacity and Minimum Dowel Area

Column Size	φP_{nb}, (kips)	Minimum dowel area (A_s = 0.005 A_g)
8x8	152^k	0.32 in^2
10x10	238	0.50
12x12	343	0.72
14x14	467	0.98
16x16	609	1.28
18x18	771	1.62
20x20	952	2.00
22x22	1152	2.42
24x24	1371	2.88

When the factored column load P_u exceeds the concrete bearing capacity φP_{nb}, the excess compression must be transferred to the footing by footing dowel bars. Dowel bars can be the same size as or smaller than the column bars, but cannot be more than one bar size larger than the column bars (ACI 15.8.2.3). Total area of dowel bars cannot be less than 0.5% of the column } * area (ACI 15.8.2.1); see Table 7-4. A minimum of 4 dowels is recommended (one in each corner of the column).

Total length of dowel = embedment in footing + lap splice in column.

See Fig. 7-2. Both embedment length in footing and lap splice in column must develop the larger of two conditions:

(1) Excess compression (if any) transferred by dowels, or

(2) One-fourth the tensile capacity of the column bars in each column face (ACI 12.17.3).

Condition (1) requires dowel development in compression and Condition (2), dowel development in tension.

For dowel embedment length into footing, either straight bar embedment or standard 90° hooked-bar embedment can be used.

Minimum dowel bar embedment lengths are tabulated in Table 7-5. The larger of the compression development ℓ_d and the tension development, either straight-bar ℓ_d or hooked-bar ℓ_{dh}, must be used to satisfy both development conditions (1) and (2). For compression development, only the straight vertical portion of the dowel is considered effective. Tabulated embedment lengths may be reduced to account for excess dowel area provided... A_s(req'd)/A_s(prov'd).

For tension development [Condition (2)], the 90°-hooked dowel bar requires less overall embedment, especially for the larger dowel bar sizes; the hook detail will also allow for easier placement and attachment of the footing dowels directly on top of the footing reinforcement by bar ties, prior to casting the footing concrete.

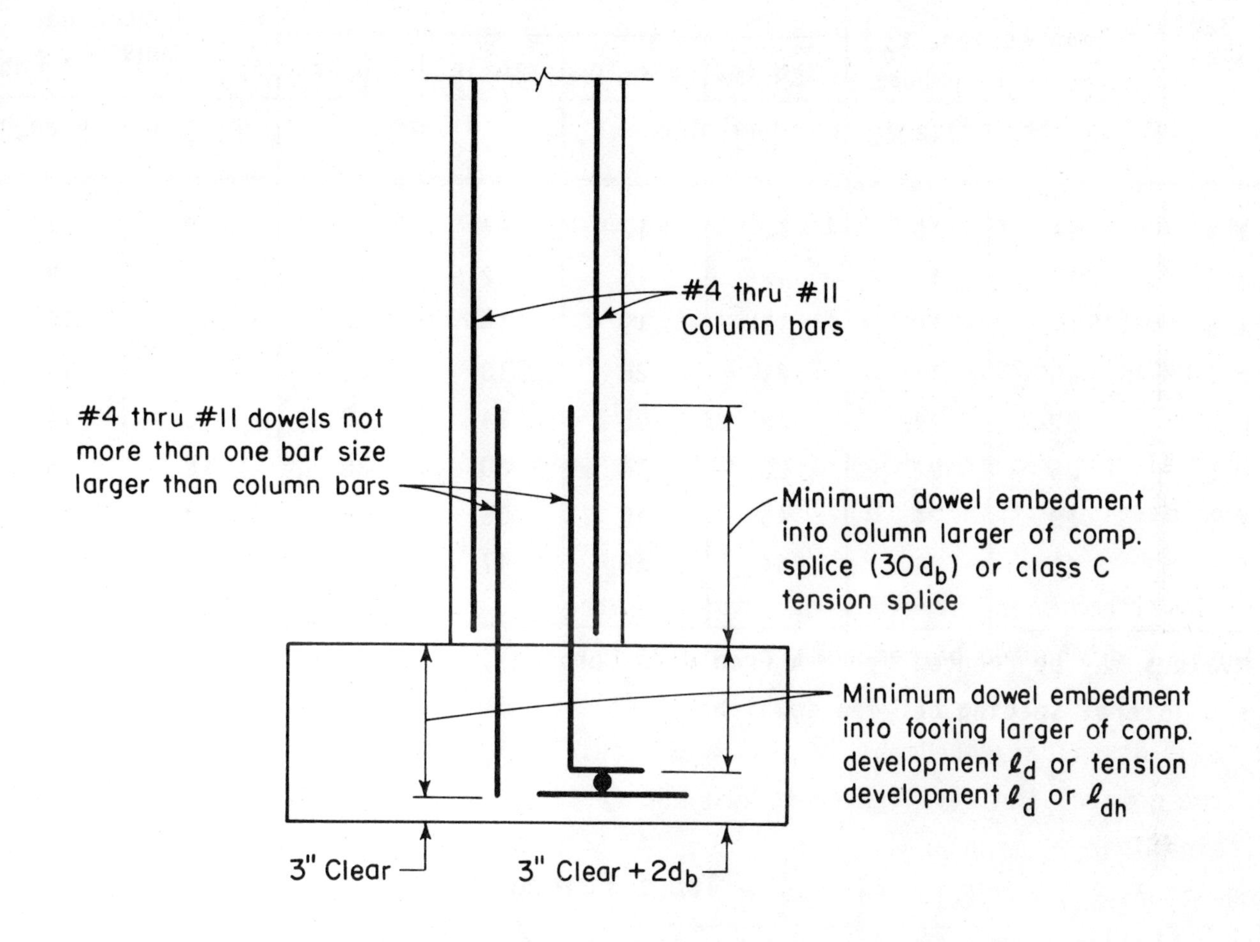

Fig. 7-2 Footing Dowels

Table 7-5 - Minimum Dowel Bar Embedment into Footing* (inches)

Dowel Bar Size	Condition (1)		Condition (2)					
	Compression, ℓ_d		Straight-bar tension, ℓ_d				hooked-bar tension, ℓ_{dh}	
			s≥6 in.	s<6 in.	s≥6 in.	s<6 in.		
	f'_c=3000	f'_c=4000	f'_c=3000		f'_c=4000		f'_c=3000	f'_c=4000
# 4	11	10	12	12	12	12	8	7
# 5	14	12	12	15	12	15	10	9
# 6	17	15	15	19	14	18	12	10
# 7	19	17	21	26	18	23	14	12
# 8	22	19	28	35	24	30	16	14
# 9	25	22	35	44	30	38	18	15
#10	28	24	45	56	38	48	20	17
#11	31	27	54	68	47	59	22	19

*Values may be reduced by....A_s(req'd)/A_s(prov'd).

s = lateral spacing between dowel bars.

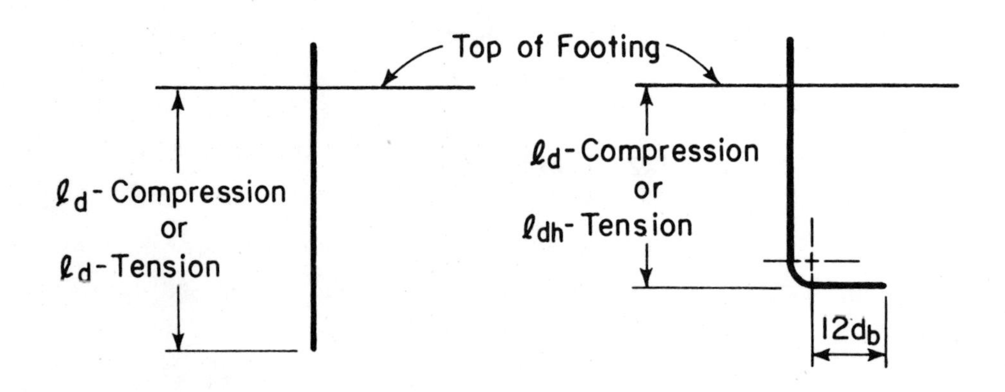

Straight-dowel detail

Hooked-dowel detail

Alternatively, straight bar embedment into the footing, especially for plain concrete footings, may expedite construction by first casting the footing concrete, then embedding the footing dowels while the concrete is still in a plastic state. This construction practice is specifically permitted by ACI 16.4.2. The dowels must, however, be maintained in correct position while concrete remains plastic. Before giving approval, the engineer should be satisfied that the detailing and workmanship will result in properly placed dowels and that the concrete will be properly compacted around the dowels.

For minimum footing thickness with straight dowel bars, add 3 in. cover to required dowel embedment length. For minimum footing thickness with hooked dowel bars, add 5 in. (3 in. cover + 2 bar dia. ≃ 2 in.) to required dowel embedment length. If required length of dowel embedment into footing is greater than footing thickness, three possibilities are open:

1. Add a monolithic concrete cap sufficiently thick to provide total depth required for dowel embedment.

2. Use a greater number of smaller size dowels to provide required dowel area and required dowel embedment length within footing thickness.

3. Thicken footing to provide necessary dowel embedment.

Embedment length for dowels into the column must also develop both compression and tension capacity by proper splice length with the column vertical bars. Minimum dowel embedment (splice) lengths into the column are tabulated in Table 7-6. The larger of the compression and tension spliced lengths must be used to satisfy both development conditions. Splice length is based on size of dowel bar.

Table 7-6 - Minimum Straight Dowel Bar Embedment into Column* - inches

Dowel Bar Size	Condition (1)	Condition (2)	
	Compression Splice	Tension Splice	
		s ≥ 6 in.	s < 6 in.
# 4	15	16	20
# 5	19	20	26
# 6	23	24	31
# 7	27	31	39
# 8	30	41	51
# 9	34	51	65
#10	38	65	82
#11	43	80	100

*Values may be reduced by...A_s(req'd)/A_s(prov'd)
s = lateral spacing between dowel bars.
f'_c = 4000 psi

7.6.2 - Horizontal Force Transfer at Base of Column

Footing dowels may be required to transfer horizontal load from wind, earth pressure, etc. at base of column to footing. For this design condition, the shear-friction design method (ACI 11.7) provides a convenient design tool for checking the horizontal load-transfer strength of the footing dowels. The dowels provided for vertical force transfer may also act as horizontal load-transfer reinforcement; the dowels need only be adequate for the more severe of the two design conditions. The required area of footing dowels to transfer a horizontal force H_u is computed directly from the shear-friction design equation:

$$A_s = \frac{H_u}{\varphi f_y \mu}$$

Where μ = 0.6 for concrete surface not intentionally roughened.

μ = 1.0 for footing concrete in contact with the column concrete intentionally roughened by heavy raking or grooving (ACI 11.7.4.3). With the intentionally roughened surface, a 40% reduction in required A_s results.

φ = 0.85, as for shear.

Note: The horizontal force H_u to be transferred cannot exceed $\varphi(0.2\ f'_c A_c)$, or for our column concrete strength f'_c = 4000 psi, $\varphi(800\ A_c)$. A_c is gross area of column.

Dowels required to transfer horizontal force must be anchored for tension, with tension development into the footing (Table 7-5) and tension splice length into the column (Table 7-6).

7.7 EXAMPLE: REINFORCED FOOTING DESIGN

Design footings for the interior columns of Bldg. #2 (5-Story Flat Plate). Assume base of footings located 5 ft below first story floor slab, with sand and clay footing bed. Permissible soil pressure q_a = 4.5 ksf.

(1) Design Data:

Service surcharge = 50 psf
Assume weight of soil and concrete above footing base = 130 pcf.

Interior columns - 16 in. x 16 in.
4 #8 bars (braced frame)
8 #10 bars (unbraced frame)

f'_c = 4000 psi (column)
f'_c = 3000 psi (footing)

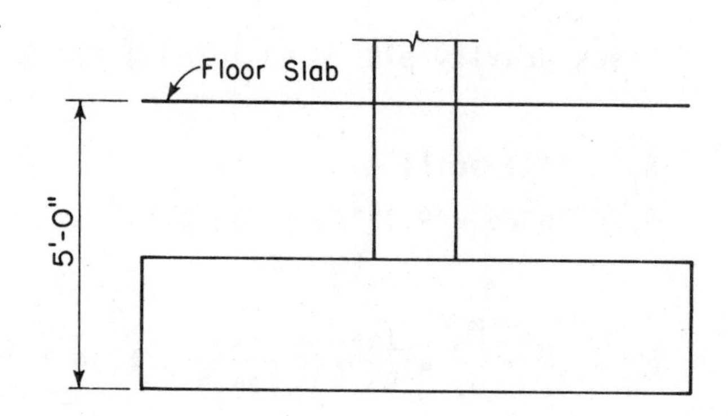

(2) Load combinations to be considered (see Example 5.6.1)

(a) gravity loads: $P_u = 576^k$ Eq. (9-1)

$M_u = 11'^k$

(b) gravity + wind loads: $P_u = 432^k$ Eq. (9-2)

$M_u = 129(1.34) = 173'^{k*}$

*For the unbraced frame, total moment to be transferred to the footing is equal to the magnified column moment.

(3) Base area of footing

Determine footing base area for gravity loads only, then check footing size for gravity plus wind loads.

Total weight of surcharge = 0.130 x 5 + 0.05 = 0.70 ksf
Net permissible soil pressure = 4.5 - 0.70 = 3.8 ksf

$$A_f{}^* = \frac{P_u}{1.6q_a} = \frac{576}{1.6(3.8)} = 94.7 \text{ sq ft}$$

*Neglect small gravity load moment for sizing footing

<u>Use 9 ft-9 in. x 9 ft-9in. square footing</u>
(A_f = 95.1 sq ft)

Check gravity plus wind loading for 9 ft-9 in. x 9 ft-9 in. footing.

A_f = 95.1 sq ft
$S_f = bh^2/6 = 9.75^3/6 = 154.5 \text{ ft}^3$

$$q_u = \frac{P_u}{A_f} + \frac{M_u}{S_f} = \frac{432}{95.1} + \frac{173}{154.5} = 4.54 + 1.12 = 5.66 \text{ ksf}$$

$< 1.6(3.8) = 6.08$ ksf OK

Note: With gravity loads governing, the interior footings are the same for both the braced and unbraced frame alternatives.

(4) Footing thickness

Footing projection = (9.75 - 16/12)/2 = 4.21 ft

$$h = 2.2\sqrt{\frac{P_u c^2}{A_f}} + 4 \text{ in.} = 2.2\sqrt{\frac{576(4.21)^2}{95.1}} + 4 = 22.8 + 4 = 26.8 \text{ in.}$$

<u>Use h = 27 in. (2 ft-3 in.)</u>

(5) Footing reinforcement

$A_s = 0.022\, h = 0.022(27) = 0.594 \text{ in.}^2/\text{ft}$

possible bar selection (Table 7-2): #6 @ 9 in. ($A_s = 0.59 \text{ in.}^2/\text{ft}$)
#7 @ 12 in. ($A_s = 0.60 \text{ in.}^2/\text{ft}$)
#8 @ 16 in. ($A_s = 0.59 \text{ in.}^2/\text{ft}$)

Check available ℓ_d for bar development (Table 7-3):

For $f'_c = 3000$:

#6 bars, $\ell_f \geq 2(15) + 16 + 6 = 52/12 = 4.33$ ft < 9.75 ft OK

#7 bars, $\ell_f \geq 2(21) + 22 = 64/12 = 5.33$ ft < 9.75 ft OK

#8 bars, $\ell_f \geq 2(28) + 22 = 78/12 = 6.5$ ft < 9.75 ft OK

Select #7 @ 12, total bars required = (117 - 6)/12 = 9.25 spaces

<u>Use 10 #7 bars x 9 ft-3 in. (each way)</u>

(6) Footing dowels

Footing dowel requirements will differ for the braced and unbraced frame. For the unbraced frame, with substantial wind moment transfer at base of column, provide equal size dowel bars for the total column bars. Use 8 #10 dowel bars, and embed for full tension development into both column and footing.

For the braced frame, subject to gravity loads only, dowel requirements are determined as follows.

(a) For 16 x 16 column (Table 7-4):

$\varphi P_{nb} = 609^k$

$A_s(\min) = 1.28 \text{ in.}^2$

$\varphi P_{nb} > P_u$

609 > 576 Bearing on concrete alone is adequate for transfer of compressive force. Dowel lengths need only satisfy Condition (2)--minimum tensile capacity.

Area of dowels required for minimum tensile capacity (ACI 12.17.3):

A_s (2 #8 column bars)/4 = (2 x 0.79)/4 = 0.395 in.^{2}

For 4 dowels, 1.28/4 = 0.32 in.^{2}/dowel x 2 = 0.64 > 0.395

<u>Use 4 #6 dowels (A_s = 1.76 in.^{2})</u>

(b) Embedment into footing (Table 7-5):

For straight dowel bars, development depth available =

27 - 3 = 24 in.

For #6 dowels
f'_c = 3000 psi ⎤ ℓ_d = 15 in.
s > 6 in.

ℓ_d can be reduced by A_s(req'd)/A_s(prov'd) = 0.395/2 x 0.44 = 0.45

ℓ_d = 15(0.45) = 6.75 Use 12 in. (min.)

For hooked dowel bars, development depth available = 27 - 5 = 22 in.

For #6 dowels
f'_c = 3000 psi] ℓ_{dh} = 12 in. (min.)

If hooked dowel bars are preferred for easier placement directly on top of the footing reinforcement, use 22 in. embedment length into footing.

(c) Embedment into column (Table 7-6):

For #6 dowels
f'_c = 4000 psi
s > 6 in.] splice length = 24 in.

For excess area = 24(0.45) = 10.8 in. Use 12 in. (min.)

Use 4 #6 dowels x 2 ft-9 in. with standard 90° end-hook

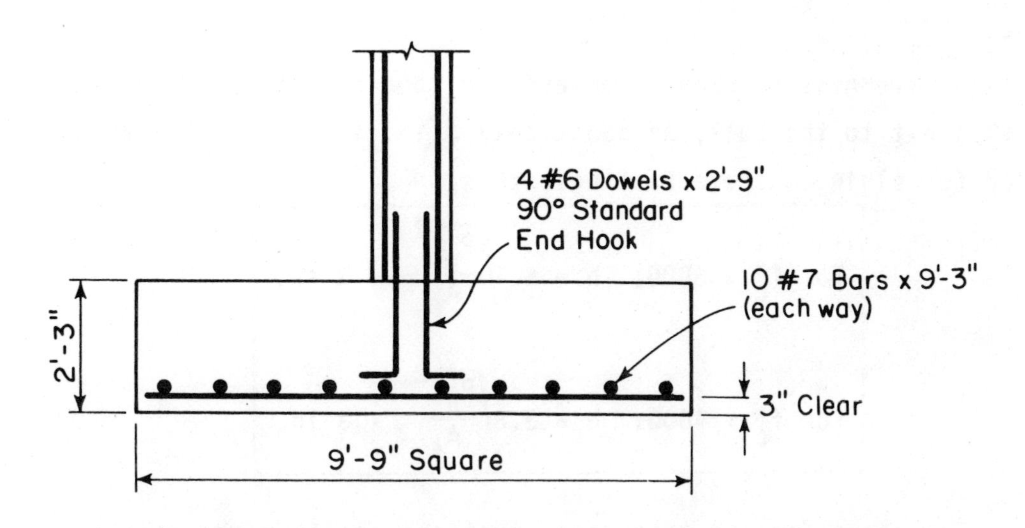

Reinforced Footing Details

7.8 ONE-STEP THICKNESS DESIGN FOR PLAIN FOOTINGS

Depending on magnitude of column or wall loads and site soil conditions, plain concrete footings may be an economical alternative to reinforced concrete footings. As for reinforced footings, a simplified one-step thickness design equation can be derived for plain footings. Structural plain concrete members are designed according to ACI Standard 318.1.[7.2] For plain concrete, the maximum permissible flexural tension stress under factored load conditions is limited to $5\varphi\sqrt{f'_c}$ (ACI 318.1 - 6.2.1). For a φ factor of 0.65 (ACI 318.1 - 6.2.2), the permissible tension f_t is,

For f'_c = 3000 psi; f_t = 178 psi
For f'_c = 4000 psi; f_t = 206 psi

For design per foot width of footing, referring to Fig. 7-3:

$$M_u \leq M_n$$

$$q_s \left(\frac{c^2}{2}\right) \leq f_t \left(\frac{h^2}{6}\right)$$

$$h^2(\text{req'd}) = 3q_s \frac{c^2}{f_t} = \frac{3P_u c^2}{A_f f_t}$$

To allow for unevenness of excavation and for some contamination of the concrete adjacent to the soil, an additional 3 in. in overall thickness is recommended for plain concrete footings; thus,

$$\text{For } f'_c = 3000\text{:} \quad h = 4.1\sqrt{\frac{P_u c^2}{A_f}} + 3 \text{ in.}$$

$$\text{For } f'_c = 4000\text{:} \quad h = 3.8\sqrt{\frac{P_u c^2}{A_f}} + 3 \text{ in.}$$

Note: The above footing thickness equations are in mixed units;
P_u = factored column load, kips
A_f = base area of footing, sq ft

c = greatest distance from face of column to edge of footing, ft

h = overall thickness of footing, in. ≥ 8 in. (ACI 318.1 - 7.2.4)

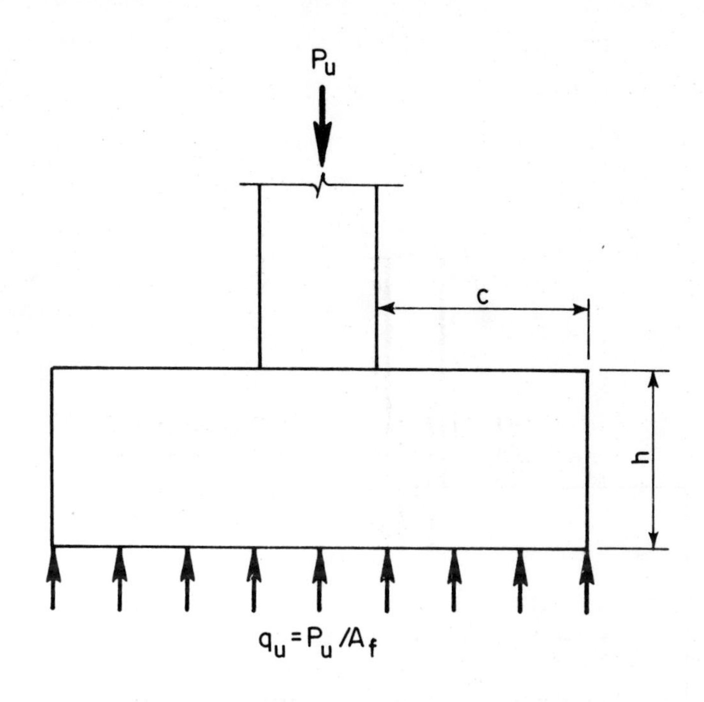

7-3 - Plain Footings

Comment on Shear Strength--Thickness of plain concrete footings will be controlled by flexural strength rather than shear strength for the usual proportions of plain concrete footings. Shear rarely will control.

7.8.1 - Example: Plain Footing Design

For the interior columns of Bldg. #2, Alternate (2)-with Structural Walls, provide a plain footing alternative.

From Example 8.7:

A_f = 9 ft-9 in. x 9 ft-9 in. = 95.1 sq ft

$P_u = 576^k$

c = footing projection = 4.21 ft

For f'_c = 3000 psi (footing concrete):

$$h = 4.1\sqrt{\frac{P_u c^2}{A_f}} + 3 \text{ in.} = 4.1\sqrt{\frac{576(4.21)^2}{95.1}} + 3 = 42.5 + 3 = 45.5 \text{ in.}$$

<u>Use h = 45 in. (3 ft-9 in.)</u>

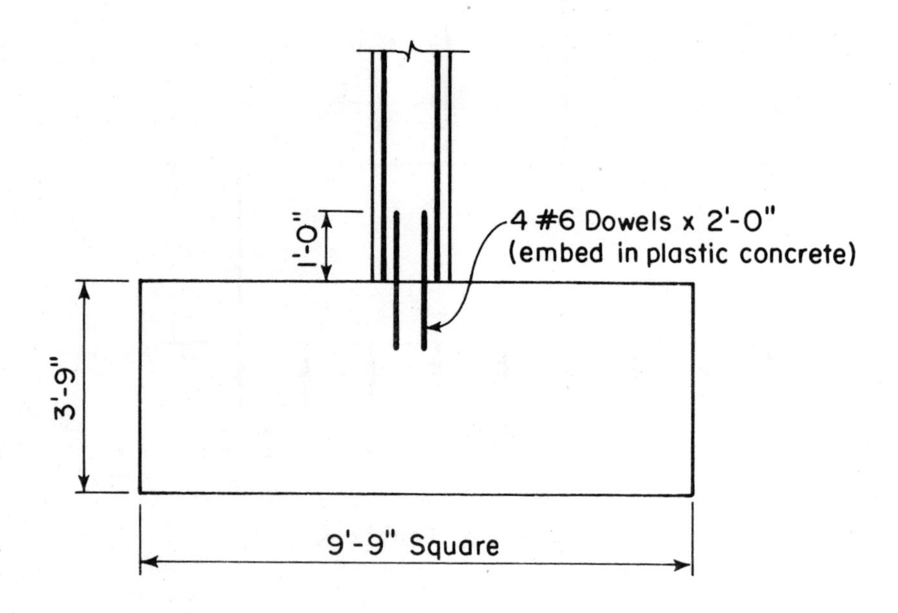

Plain Footing Alternate

Selected References

7.1 "Notes on ACI-83, Chapter 9, Design for Flexure," 4th Edition, EB070.04D, Portland Cement Association, Skokie, Ill.

7.2 "Building Code Requirements for Structural Plain Concrete (ACI 318.1-83)," American Concrete Institute, Detroit, 1983, 7 pp.

8

Structural Detailing of Reinforcement ...for Economy

David P. Gustafson*

8.1 - INTRODUCTION

In reinforced concrete, more of the "art" of structural design ought to be directed toward reinforcement details and arrangements of reinforcing bars. There are at least three interrelated requirements that should be considered in this part of the "art" of design, since structurally sound details and proper bar arrangements....

- Are vital to the satisfactory performance of reinforced concrete structures
- Should be practical and buildable
- Should be cost-effective

Ideally, the economics of reinforced concrete should be viewed in the broad perspective, considering all facets in the execution of a project. While it may be important to strive for savings in materials, many engineers often tend to focus too much on material savings rather than designing for construction efficiencies. No doubt a savings in material quantities should result from a highly refined or, literally, a custom design for each structural member in a building. However, such a savings in materials might be false economy if significantly higher construction costs are incurred in building the "custom-designed" members.

*Technical Director, Concrete Reinforcing Steel Institute.

Use of the concept of "trade-offs" should be considered in holding or reducing the total cost of construction, including the total in-place cost of reinforcement. Savings in reinforcement weight can be "traded-off" for cost savings in fabrication, placing, and inspection for overall economy.

8.2 - DESIGN CONSIDERATIONS FOR REINFORCEMENT ECONOMY

The following notes on reinforcement selection and placement will provide for overall economy and will minimize costly project delays and job stoppages:

(1) First and foremost, **show clear and complete reinforcement details and bar arrangements** in the Contract Documents. This issue is addressed in Reference 8.1: "...the responsibility of the Engineer is to furnish a clear statement of design requirements; the responsibility of the [Reinforcing Steel] Detailer is to carry out these requirements."

The ACI Code further emphasizes that the designer is responsible for the types and locations of splices of reinforcement [ACI 1.2.1(h) and 12.14.1].

(2) Use **Grade 60** reinforcing bars. The apparent savings in tonnage with Grade 60 bars versus Grade 40 bars is about 33%. However, due to longer development lengths and lap splice lengths for Grade 60 bars and Code provisions for minimum steel areas, limits on maximum spacings, distribution of flexural reinforcement, etc., the optimum savings realistically will be somewhat less. In building structures, the achievable savings in tonnage is in order of 20 to 25%.

Use bar sizes #11 and smaller; #14 and #18 bars are not generally inventoried in regular stock. Also, bar sizes smaller than #6 generally cost more per pound and require more placing labor per pound of reinforcement.

(3) Use straight bars only in flexural members. Straight bar designs are regarded as standard in the industry. Truss-bent bars are undesirable from a fabrication and placing standpoint, and structurally unsound where stress reversals occur.

(4) In beams, specify bars in single layer only. Use one bar size for reinforcement in one face at a given location in a span. In slabs, space reinforcement in whole inches, but not less than 6-inch spacing.

(5) Use largest bar sizes possible for the longitudinal reinforcement in columns. Use of larger bar sizes and fewer bars in other structural members will be restricted by Code requirements for development of reinforcement, limits on maximum spacings, and distribution of flexural reinforcement.

(6) Use or specify fewest possible bar sizes for a project.

(7) Stirrups are typically the smaller bar sizes, which usually results in the highest total in-place cost of reinforcement per ton. For overall economy and to minimize congestion of reinforcement, specify the largest size and fewest number of stirrups and the fewest variations in spacing.

(8) When closed stirrups are required, specify practical two-piece closed types to facilitate placing.

(9) Fit and clearance of reinforcing bars warrant special attention by the Engineer. At beam-column joints, arrangement of column bars must provide enough space or spaces to permit passage of beam bars. Bar details should be properly prepared and reconciled before the bars are fabricated and delivered to the job site. Member connections are far too important to require indiscriminate adjustments in the field to facilitate bar placing.

(10) Use or specify **standard reinforcing bar details and practices:**

- Standard end hooks (ACI 7.1); note that the provisions in ACI 12.5 are only applicable for determining the anchorage capacity of standard hooks conforming to ACI 7.1.

- Typical bar bends (Fig. 6 in Reference 8.1).

- Standard fabricating tolerances (Figs. 4 and 5 in Reference 8.1); more restrictive tolerances must be indicated by the Engineer in the Contract Documents.

- Tolerances for placing reinforcing bars (ACI 7.5); more restrictive tolerances must be indicated by the Engineer in the Contract Documents.

- Care must be exercised in specifying more restrictive tolerances for fabricating and placing reinforcing bars. More restrictive fabricating tolerances are limited by the capabilities of shop fabrication equipment. Fabricating and placing tolerances must be coordinated. Tolerances for the formwork must also be considered and coordinated.

(11) Never permit **field welding** of crossing reinforcing bars for assembly of reinforcement ("tack" welding, "spot" welding, etc.). Tie wire will do the job without harm to the bars.

(12) Avoid **manual arc-welded** splices of reinforcing bars in the field wherever possible, particularly for smaller projects.

(13) A frequently occurring construction problem is having to make **field corrections** to reinforcing bars partially embedded in hardened concrete. Such "job stoppers" usually result from errors in placing or fabrication, accidental bending caused by construction equipment, or

a design change. Field bending of bars partially embedded in concrete is not permitted except if such bending is shown on the design drawings or authorized by the Engineer (ACI 7.3.2). The ACI Code Commentary offers guidance on rebending with the use of heat. Further guidance on bending and straightening of reinforcing bars is reported in Reference 8.2.

8.3 - REINFORCING BARS

Billet-steel reinforcing bars conforming to ASTM A 615, Grade 60, are the most widely used type and grade in the United States. Combining the Strength Design Method with Grade 60 bars results in maximum overall economy. This design practice has made Grade 60 reinforcing bars the standard grade. The current edition of ASTM A 615 reflects this practice as only bar sizes #3 through #6 in Grade 40 are included in the specification.

When important or extensive welding is required, or when more bendability and controlled ductility are required as in seismic-resistant design, use of low-alloy reinforcing bars conforming to ASTM A 706 should be considered. However, local availability should be investigated before specifying A 706 bars.

8.3.1 - Coated Reinforcing Bars

Zinc-coated (galvanized) and epoxy-coated reinforcing bars are used increasingly as a corrosion-protection system in reinforced concrete structures. An example of a structure that might use coated bars is a parking garage where vehicles track in deicing salts.

Zinc-coated (galvanized) reinforcing bars must conform to ASTM A 767, and the coated bars must conform to ACI 3.5.3.1. Bars are usually galvanized after fabrication. The Engineer should specify which bars require special finished bend diameters, usually the smaller bar sizes for stirrups and ties. ASTM A 767 has two classes of coating weights. Class I (3.5 oz/sq ft) is normally specified for general construction. ASTM A 767 contains three supplementary requirements S1, S2, and S3. S1 requires sheared ends to be coated with a zinc-rich formulation. When bars are fabricated after galvanizing, S2 requires damaged coating to be repaired with a zinc-rich formulation. If ASTM A 615 billet-steel bars are being supplied, S3 requires that a silicon analysis of each heat of steel be provided. It is recommended that S1 and S2 be specified when bars are fabricated after galvanizing.

Uncoated reinforcing steel (or any other embedded metal dissimilar to zinc) should not be permitted in the same concrete element with galvanized bars, nor in close proximity to galvanized bars, except as part of a cathodic-protection system. Galvanized bars should not be coupled to uncoated bars.

Epoxy-coated reinforcing bars must conform to ASTM A 775, and the bars that are to be coated must conform to ACI 3.5.3.1.

Proper use of ASTM A 767 and A 775 requires the inclusion of provisions in the project specifications for the following items:

- Compatible tie wire, bar supports, support bars, and spreader bars in walls.
- Requirements for repair of damaged coating after completion of welding (splices) or installation of mechanical connections.
- Requirements for repair of damaged coating after completion of field corrections, when field bending of coated bars partially embedded in concrete is permitted.
- Requirements to minimize damage to coated bars during handling, shipment, and placing operations and limits on permissible coating damage and, when required, repair of damaged coating.

Reference 8.3 contains suggested provisions for the preceding items for epoxy-coated reinforcing bars.

8.4 - DEVELOPMENT OF REINFORCING BARS

The fundamental requirement for development or anchorage of reinforcing bars is that a reinforcing bar must be embedded in concrete a sufficient distance on each side of a critical section to develop the calculated tension or compression in the bar at the section. The Code defines "development length" as the length of an embedded reinforcing bar required to develop the design strength of the bar at a critical section. Standard end hooks (or mechanical devices) may be used for anchorage of reinforcing bars, except that hooks are only effective for developing bars in tension (ACI 12.1).

8.4.1 - Development of Straight Bars in Tension

Tension development length ℓ_d for deformed bars (ACI 12.2) includes consideration of a number of modification factors that either increase or decrease the "basic" development length ℓ_{db}. For our selected materials, normal weight concrete with Grade 60 bars, the only modification factor that results in an increase in the basic development length is for the "top bar" effect (ACI 12.2.3.1). Development length for bars classified as "top bars" must be increased by 40%. This rather substantial increase reflects the condition that "top bars" may have reduced anchorage due to settlement of the concrete below the bars. It should be noted that "top bars" are defined in the Code as horizontal bars so placed that more than 12 in. of fresh concrete is cast in the member below the bars. Reinforcing bars physically located in the top part of a relatively shallow structural member may not necessarily be "top bars" for development length considerations. For example, the top layer of bars in a slab would be classified as "top bars" only if the overall slab thickness is at least equal to the sum, h=12 in. + d_b + cover. For #6 bars with 3/4-in. cover, h would have to be at least 13-1/2 in. in order for the bars to be defined as "top bars."

The basic development length may be reduced 20% if the bars are spaced laterally at least 6 in. on center with at least 3 in. side cover in the plane of the bars (ACI 12.2.4.1). Basic development length may also be reduced if an excess area of steel is provided (ACI 12.2.4.2).

For design convenience, tension development lengths for Grade 60 rebars are tabulated in Table 8-1. The modification factors for "top bar" (1.4) and for wider bar spacing (0.8) are included in the tabulated values. The development lengths for ($s < 6$ in.) are for general use where spacing of bars being developed is less than 6 in. on center. The ($s \geq 6$ in.) lengths apply where lateral spacing between bars is 6 in. or greater on center; edge bar must also be spaced with at least 3 in. clear cover to edge of member. Where development for full f_y is not specifically required, the tabulated values may be reduced for excess reinforcement by the factor [A_s req'd/A_s prov'd], but not less than 12 in. (ACI 12.2.5).

Table 8-1 Minimum Straight Tension Development Length ℓ_d (inches) for Grade 60 Rebars

Bar Size	f'_c = 3000 psi*				f'_c = 4000 psi*			
	($s < 6$ in.)		($s \geq 6$ in.)		($s < 6$ in.)		($s \geq 6$ in.)	
	Top Bars	Other Bars	Top Bars	Other Bars	Top Bars	Other Bars	Top Bars	Other Bars
#3	13	12	12	12	13	12	12	12
#4	17	12	13	12	17	12	13	12
#5	21	15	17	12	21	15	17	12
#6	27	19	22	15	25	18	20	14
#7	37	26	29	21	32	23	26	18
#8	48	35	39	~~88~~ 28	42	30	34	24
#9	61	44	49	35	53	38	43	30
#10	78	56	62	45	67	48	54	39
#11	96	68	77	55	83	59	66	47

*Normal weight concrete

8.4.2 - Development of Hooked Bars in Tension

Development length ℓ_{dh} for deformed bars terminating in a standard hook (ACI 12.5) are tabulated in Table 8-2 for Grade 60 Rebars. Similar to straight bar development, hooked bar development ℓ_{dh} includes consideration of modification factors that either increase or decrease the "basic" hook development length ℓ_{hb}. The modification factors of ACI 12.5.3.2 and 12.5.3.3 are pertinent; both factors can provide significantly shorter hooked bar embedment lengths. The 30% and 20% reduction factors, respectively, account for the favorable confinement conditions provided by increased concrete cover and/or transverse ties or stirrups to resist splitting of the concrete.

Embedment lengths given in Table 8-2(a) General Use, apply for end hooks with maximum side cover normal to plane of hook of 2-1/2 in. and minimum end cover (90° hooks only) of 2 in.; $\ell_{dh} = (0.7)\ell_{hb}$, but not less than $8d_b$ nor 6 in. Note: for hooked bar anchorage in beam-column joints, the beam hooked bars are usually placed inside the column vertical bars with side cover greater than the 2-1/2-in. minimum required for application of the 0.7 factor. Also, for 90° end hooks with hook extension located inside the column ties, the 2-in. minimum will usually be satisfied to permit the 0.7 reduction factor.

The special confinement condition given in Table 8-2(b) includes the additional 0.8 reduction factor for confining ties or stirrups; $\ell_{dh} = (0.7\text{x}0.8)\ \ell_{hb}$, but not less than the absolute minimum $8d_b$ or 6 in.

Where development for full f_y is not specifically required, the tabulated values of Table 8-2 may be further reduced for excess reinforcement provided [A_s req'd/A_s prov'd], but not less than $8d_b$ nor 6 in.

Note: The resulting ℓ_{dh} dimension should be shown on the design drawings.

One additional consideration for hooked bar anchorage....hooked bars terminating at the discontinuous end of members (ends of simply supported beams, free end of cantilevers, and ends of members framing into a joint where the member does not extend beyond the joint) must satisfy the additional confinement required by ACI 12.5.4. If the full strength of the hooked bar must be developed, and if both the side cover and top (or bottom) cover over the hook is less than 2-1/2 in., closed ties or stirrups spaced at $3d_b$ are required over the development length ℓ_{dh}. At discontinuous ends of slabs with concrete confinement provided by the slab continuous on both sides normal to the plane of the hook, the requirements for confining ties or stirrups of ACI 12.5.4 do not apply.

Table 8-2 Minimum Embedment Length ℓ_{dh} (inches) for Standard End Hooks on Grade 60 Bars

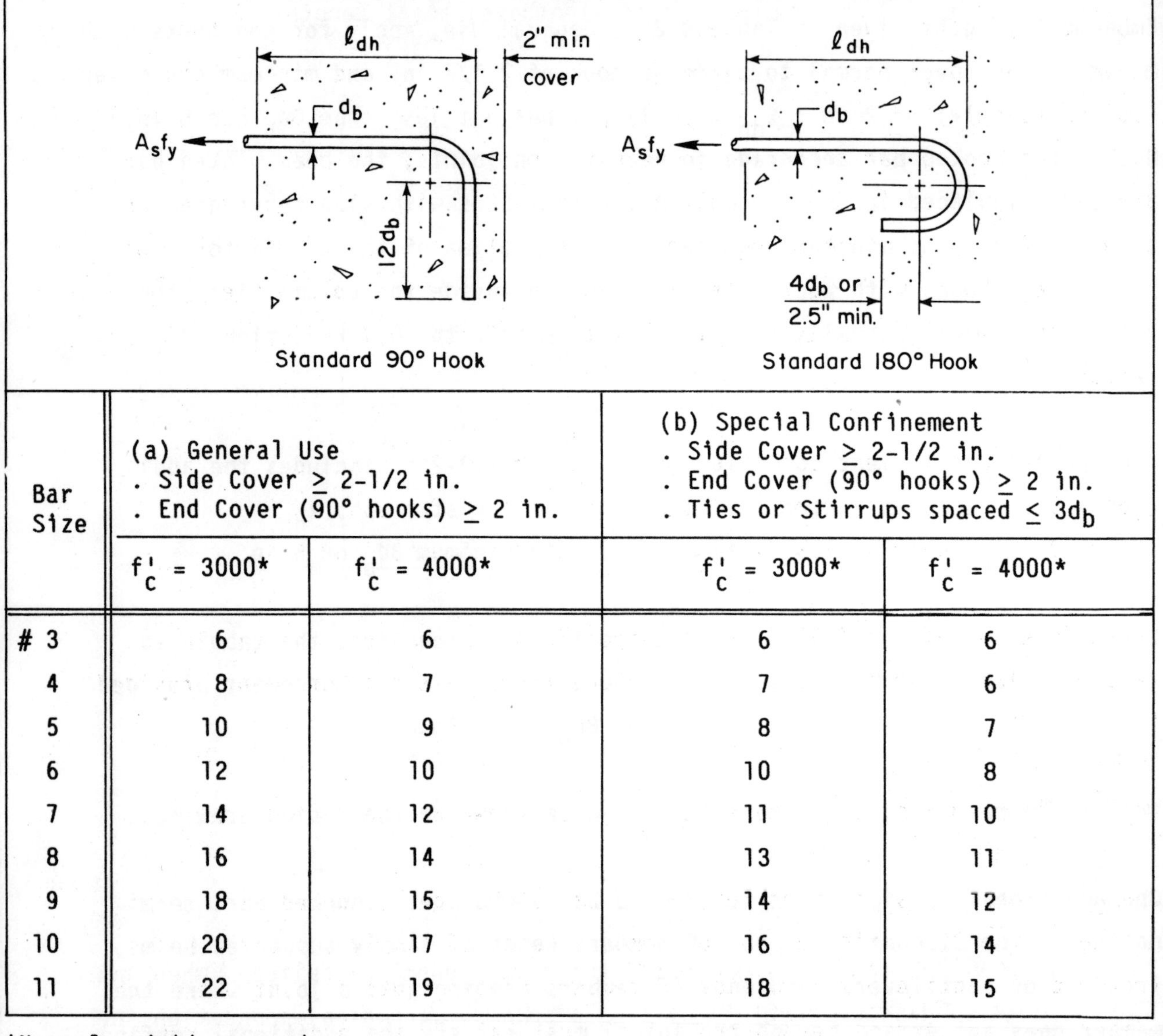

Standard 90° Hook

Standard 180° Hook

Bar Size	(a) General Use . Side Cover ≥ 2-1/2 in. . End Cover (90° hooks) ≥ 2 in.		(b) Special Confinement . Side Cover ≥ 2-1/2 in. . End Cover (90° hooks) ≥ 2 in. . Ties or Stirrups spaced ≤ $3d_b$	
	f'_c = 3000*	f'_c = 4000*	f'_c = 3000*	f'_c = 4000*
# 3	6	6	6	6
4	8	7	7	6
5	10	9	8	7
6	12	10	10	8
7	14	12	11	10
8	16	14	13	11
9	18	15	14	12
10	20	17	16	14
11	22	19	18	15

*Normal weight concrete

8.4.3 - Development of Bars in Compression

Compression development lengths (ACI 12.3) for Grade 60 rebars are tabulated in Table 8-3. The values apply for all concrete strengths equal to or greater than 3000 psi.

Table 8-3 Minimum Compression Development Lengths ℓ_d (inches) for Grade 60 Bars ($f'_c \geq 3000$ psi)

Bar Size	ℓ_d*
#3	8
#4	11
#5	14
#6	17
#7	19
#8	22
#9	25
#10	28
#11	31

*Not less than 8 in.

8.5 - SPLICES OF REINFORCING BARS

Three methods are used for splicing reinforcing bars:

- •Lap splices
- •Welded splices
- •Mechanical connections

The traditional lap splice is generally the most economical splice. When lap splices cause congestion or field placing problems, mechanical connections or welded splices should be considered. The location of construction joints, provision for future construction, and the particular method of construction can also make lap splices impractical. Also, in columns, lapped offset bars may need to be located inside the bars above to reduce reinforcement congestion, thus reducing the moment capacity of the column-section at the lapped splice location. When the amount of column vertical reinforcement is greater than 4%, and particularly in combination with large design moments, use of butt splices -- either mechanical connections or welded splices -- should be considered to reduce congestion and to provide for greater design moment strength of the column section at the splice locations.

Bars in lap splices may be spaced or in contact (ACI 12.14.2.3); however, contact lap splices are preferred for the practical reason that when the bars are wired together, they are more easily secured against displacement during placing of concrete.

Welded splices generally require the most expensive field labor. For projects of all sizes, manual arc-welded splices will usually be the most costly method of splicing due to direct and indirect costs of proper inspection.

Mechanical connections are made with proprietary splice devices. Performance information and test data should be secured directly from the manufacturers of such splice devices. Basic information about mechanical connections and the types of proprietary splice devices currently available are provided in Reference 8.4. Practical information on splicing and recommendations for the design and detailing of splices are given in Reference 8.5.

8.5.1 - Tension Lap Splices

Lap lengths for Class A, B, and C tension lap splices (ACI 12.15.1) for Grade 60 bars in normal weight concrete are given in Table 8-4 for f'_c = 3000 psi and Table 8-5 for f'_c = 4000 psi. The class of splice depends upon the magnitude of tensile stress in the bars at the splice location, and the percentage of total steel area that is to be spliced within the required lap length. The magnitude of the calculated tensile stress is determined in terms of (A_s provided/A_s required)....less than 2, or equal to or greater than 2. Actually, the tensile stress is being compared with less than or greater than one-half of the specified yield strength f_y of the bars. The lap splice lengths for (s < 6 in.) are for general use where spacing of bars being spliced is less than 6 in. on center. The (s ≥ 6 in.) lengths apply where lateral spacing between bars is 6 in. or greater on center, and side cover in the plane of the bars is at least 3 in.

Note: The Engineer must specify the required class of tension lap splice.

Table 8-4 Tension Lap Splice Lengths (inches) for Grade 60 Bars with f'_c = 3000 psi*

Bar Size	(s < 6 in.)						(s ≥ 6 in.)					
	Top Bars			Other Bars			Top Bars			Other Bars		
	A	B	C	A	B	C	A	B	C	A	B	C
#3	13	16	21	12	12	15	12	13	17	12	12	12
#4	17	22	29	12	16	20	13	17	23	12	12	16
#5	21	27	36	15	20	26	17	22	29	12	16	20
#6	27	35	46	19	25	33	22	28	37	15	20	26
#7	37	48	62	26	34	45	29	38	50	21	27	36
#8	48	63	82	35	45	59	39	50	66	28	36	47
#9	61	80	104	44	57	74	49	64	83	35	46	60
#10	78	101	132	56	72	95	62	81	106	45	58	76
#11	96	124	163	68	89	116	77	100	130	55	71	93

*Normal weight concrete

Table 8-5 Tension Lap Splice Lengths (inches) for Grade 60 Bars with f'_c = 4000 psi*

Bar Size	(s < 6 in.)						(s ≥ 6 in.)					
	Top Bars			Other Bars			Top Bars			Other Bars		
	A	B	C	A	B	C	A	B	C	A	B	C
#3	13	16	21	12	12	15	12	13	17	12	12	12
#4	17	22	29	12	16	20	13	17	23	12	12	16
#5	21	27	36	15	20	26	17	22	29	12	16	20
#6	25	33	43	18	23	31	20	26	34	14	19	24
#7	32	41	54	23	30	39	26	33	43	18	24	31
#8	42	55	71	30	39	51	34	44	58	24	31	41
#9	53	69	90	38	49	65	42	55	71	30	39	52
#10	67	88	115	48	63	82	54	70	92	38	50	65
#11	83	108	141	59	77	101	66	86	113	47	62	81

*Normal weight concrete

8.5.2 - Compression Lap Splices

Lap lengths for compression lap splices (ACI 12.16.1) for Grade 60 bars in normal weight concrete are given in Table 8-6. The values apply for all concrete strengths equal to or greater than 3000 psi. The tabulated values may be reduced (0.83 factor) for bars in tied columns if the lap splice is enclosed throughout its length by column ties (ACI 12.16.3), but the lap length must not be less than 12 in.

Table 8-6 Compression Lap Splice Lengths (inches) for Grade 60 Bars ($f'_c \geq 3000$ psi)

Bar Size	Minimum Lap Length*
#3	12
#4	15
#5	19
#6	23
#7	26
#8	30
#9	34
#10	38
#11	42

*Not less than 12 in.

8.6 - DEVELOPMENT OF FLEXURAL REINFORCEMENT

Meeting the overall requirements for development of flexural reinforcement in ACI 12.10, 12.11, and 12.12 can be a tedious and time-consuming task. These Code sections include provisions for:

- Bar extensions beyond points where reinforcement is no longer required to resist flexure.
- Termination of flexural reinforcement in tension zones.
- Minimum amount and length of embedment of positive moment reinforcement into supports.
- Limits on bar sizes for positive moment reinforcement at simple supports and at points of inflection.
- Amount and length of embedment of negative moment reinforcement beyond points of inflection.

Many of the specific requirements are interdependent, resulting in increased design time when the provisions are considered separately. To save design time and subsequent costs, it behooves the designer to consider the use of recommended bar arrangements and details, particularly for normally encountered conditions. Furthermore, as was discussed earlier in this chapter, there is potential overall cost savings in fabrication, placing, and inspection when such recommended bar details are used.

8.6.1 - Recommended Bar Length Details

Recommended bar lengths for continuous beams, one-way slabs, one-way joist construction, and two-way flat plates and flat slabs (with drops) are given in Figs 8-1 through 8-4. The bar length diagrams were developed from similar bar details presented in References 8.1 and 8.2. All recommended bar details use straight bars and meet the code requirements for development of flexural reinforcement. The bar lengths apply for uniformly distributed gravity loading only; for beams and slabs resisting lateral loads, length of reinforcement must be determined by analysis. The bar extensions are based on span moment variations developed using the moment coefficients of ACI 8.3.3 for beams and one-way slabs, and ACI 13.6 for two-way slabs. For normally encountered conditions, use of the recommended bar lengths will save considerable design time.

Note: To minimize the potential for field errors in placing, and to reduce placing and inspection time, single length top bars for the column strips of the two-way slab systems may be preferable.

8.7 - SPECIAL REBAR DETAILS AT SLAB-TO-COLUMN CONNECTIONS

When two-way slabs are supported directly by columns, as in flat plates and flat slabs, transfer of moment between slab and column takes place by a combination of flexure and eccentricity of shear. See Chapter 4, Section 4.4.1. The portion of the moment transferred by flexure is assumed to be transferred over a width of slab equal to the column width plus 1.5 times the slab thickness on either side of the column. For edge and interior columns, the effective slab width is (c+3h), and for corner columns (c+1.5h). Concentration of negative slab reinforcement is used to resist the transfer moment on this effective slab width. An effective way to achieve the concentration of reinforcement is to specify additional top bars. Note: the typical details for reinforcement of ACI Fig. 13.4.8 do not address this problem and perhaps even mislead some designers into overlooking this important detail requirement. The additional top bars must fit within the effective slab width and meet minimum bar spacing requirements. Two practical problems are immediately created for the designer. First, he must ensure that the top bars he selects for this purpose can be physically fitted into the effective slab width. Second, he must convey his design intentions clearly to both detailer and bar placer. Based on recommendations in Reference 8.7, typical details for added bars at edge and corner columns are shown in Figs. 8-5 and 8-6.

8.8 - SPECIAL SPLICE REQUIREMENTS FOR COLUMNS

8.8.1 - Construction and Placing Considerations

Of all the common structural elements, lengths of vertical bars in columns and walls are the most severely restricted. In multistory buildings, common practice is to use one-story length vertical bars in columns, erected as preassembled cages. Preassembly of column cages is usually preferred only for one-story length vertical bars, with all bars spliced at one location above each floor line.

For more heavily reinforced columns with staggered splice locations, as for butt splices with larger-size bars, vertical bars are usually in two-story lengths. The use of two-story length vertical bars will reduce the number of splices and for lap splices will result in a savings of reinforcing steel. However, the savings may be offset if more time is required to place bars in other intersecting structural members. A situation where more difficult and time-consuming placing operations might occur is in placing beam or girder bars at the intermediate floor level of the two-story length column bars. The beam or girder bars would have to be threaded through the column bars. Two-story length vertical bars might also hinder other construction operations at the job site. Most likely the projecting bars will have to be guyed for purposes of stability. The guy wires or the projecting bars may interfere with the movement of cranes for transporting construction equipment and materials.

8.8.2 - Design Considerations

Special provisions for splicing column vertical bars are contained in ACI 12.17. This section of the Code is possibly the least understood and the most misinterpreted. Determining the proper lengths of column bars, including proper splice requirements, requires knowledge of the actual stress in the bars at the critical load conditions. The required tensile strength of the spliced bars is based on the factored load stress in the bars determined for the various loading combinations that govern the design of the column (ACI 12.17.1 and 12.17.2). A minimum tensile strength of the spliced bars must be provided even when there is no calculated tension in the bars (ACI 12.17.3). Each column face must have a tensile strength at least equal to $A_s f_y/4$, where A_s is the total area of bars in each face. As an aid to the designer, the special splice requirements for columns are summarized in Table 8-7.

The designer is faced with a two-fold problem when specifying the type and practical details for column splices - - meeting the Code requirements and providing an overall cost-effective splice system. A time-saving design aid would be to have the load-moment strength diagram available for the column under consideration and, in particular, having the coordinates of five key points indicated on the diagram as illustrated in Fig. 8-7.

Table 8-7 - Special Splice Requirements for Columns

Calculated Factored Load Stress	Types of Splices Permitted	Required Tensile Strength of Splices
Ranges from f_y in compression to 1/2 f_y in tension (ACI 12.17.1)	Lap Butt welded Mechanical connections End bearing	Twice the calculated tension but not less than minimum tensile strength required by ACI 12.17.3. Provide tensile strength in each column face by splices alone or by splices in combination with continuing unspliced bars at f_y.
More than 1/2 f_y in tension (ACI 12.17.2)	Lap Full welded Full mechanical connections	Lap splices - develop f_y in tension. Full welded - develop in tension 125% f_y of bar (ACI 12.14.3.3) Full mechanical - develop in tension or compression 125% f_y of bar (ACI 12.14.4.3)

The load-moment strength diagram, drawn qualitatively, shows the Code-prescribed limits for the types of splices permitted and the required strength of splices. Using such a diagram as a design aid would involve the following steps:

(1) From the structural analysis, determine the factored loads and moments for the various load combinations.

(2) Compare the (pairs of) factored loads and moments with the coordinates of the key points and ranges of stress on the load-moment strength diagram.

(3) Observing from the preceding step which range or ranges of stress the factored loads and moments lie within provides the "ways and

means" for specifying the types and required strength of the splices.

Application of the load-moment strength diagram with the key points and ranges of stress to meet the special splice requirements is described as follows:

Point 1: $\varphi P_{n(max)}$ is the maximum design axial load strength (ACI 10.3.5); φM_n is the corresponding design moment strength.

Point 2: φP_{not} is the design axial load strength with zero computed tension; φM_{not} is the corresponding design moment strength.

Range of stress from Point 2 to Point 1 is compression only. Splice for full compression (60 ksi) and 25% tension (15 ksi) per ACI 12.17.3.

Types of splices: for #11 and smaller bars, compression lap splices, 100% of bars spliced at one section, lap lengths per Table 8-6 or end-bearing splices with 50% of bars staggered at two sections, spaced at one-half tension ℓ_d of bars (Table 8-4), provided the particular bar layout permits a division in each face to satisfy minimum 25% tension requirement (ACI 12.17.3).

Point 3: φP_{n15} is the design axial load strength at 25% computed tension (15 ksi in bars at tension face of column); φM_{n15} is the corresponding design moment strength.

Range of stress from Point 3 to Point 2 is 25% tension (15 ksi) to zero tension.

Splice for 50% minimum tension (30 ksi).

Types of splices: for #11 and smaller bars, compression lap splices, 100% of bars spliced at one section, lap lengths per Table 8-6, or end-bearing splices with 50% of bars staggered at two sections spaced at tension ℓ_d

of bars (Table 8-4) provided the particular bar layout permits a division in each face to satisfy minimum 25% of the tension requirement (ACI 12.17.3).

Point 4: φP_{n30} is the design axial load strength at 50% computed tension (30 ksi); φM_{n30} is the corresponding design moment strength.

Range of stress from Point 4 to Point 3 is 50% tension (30 ksi) to 25% tension (15 ksi).

Splice for 100% tension (60 ksi).

Types of splices: butt-welded or mechanical connections - - a minimum stagger of splices should be specified for erection purposes, or for #11 bars and smaller, tension lap splices (Tables 8-4 or 8-5); Class B if 100% of the bars are spliced at one section; Class A if 50% of the bars are spliced at one section with splices staggered at least the tension ℓ_d of the bars.

Point 5: φP_{nb} is the design axial load strength at balanced strain conditions; φM_{nb} is the corresponding design moment strength.

Range of stress from Point 5 to Point 4 is 100% tension (60 ksi) to 50% tension (30 ksi).

Types of splices: Full-welded or full mechanical connections - - a minimum stagger of splices should be specified for erection purposes, or for #11 bars and smaller, tension lap splices (Tables 8-4 or 8-5); Class C if 100% of the bars are spliced at one section; Class B if 50% of the bars are spliced at one section.

Splice details for tied columns with #11 and smaller offset bent bars are shown in Fig. 8-8. When column faces are offset more than 3 in., offset bent vertical bars are not permitted and separate dowels, lap spliced with the vertical bars, must be provided (ACI 7.8.1.5).

Typical splice details for #11 and smaller bars in tied columns with one-story and two-story length vertical bars are shown in Fig. 8-9.

Selected References

8.1 "ACI Detailing Manual - 1980" (SP-66), American Concrete Institute, Detroit, 1980.

8.2 Stecich, J. P., Hanson, J. M., Rice, P. F., "Bending and Straightening Grade 60 Reinforcing Bars," Concrete International, Vol. 6, No. 8, August 1984.

8.3 "Suggested Project Specifications Provisions for Epoxy-Coated Reinforcing Bars," Engineering Data Report No. 19, Concrete Reinforcing Steel Institute, Schaumburg, Ill., 1984.

8.4 "Mechanical Connections of Reinforcing Bars," Report No. ACI 439-3R-83, Concrete International, Vol. 5, No. 1, January 1983.

8.5 "Reinforcement Anchorages and Splices," Concrete Reinforcing Steel Institute, Schaumburg, Ill., 2nd Edition, 1984.

8.6 CRSI Handbook, Concrete Reinforcing Steel Institute, Schaumburg, Ill., 6th Edition, 1984.

8.7 "Design of Reinforcement for Two-Way Slab-to-Column Frames Laterally Braced or Not Braced," Structural Bulletin No. 9, Concrete Reinforcing Steel Institute, Schaumburg, Ill., June, 1983.

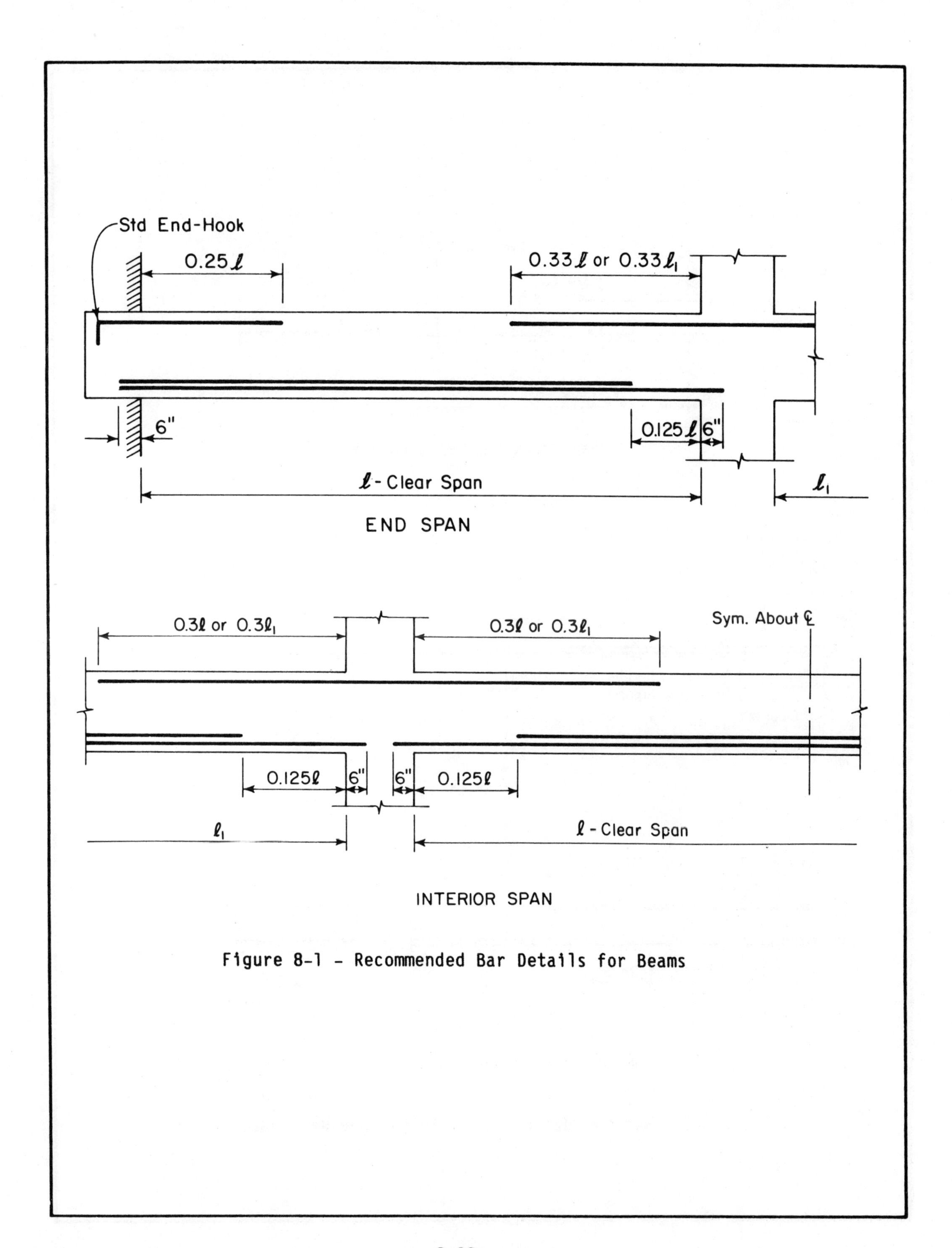

Figure 8-1 - Recommended Bar Details for Beams

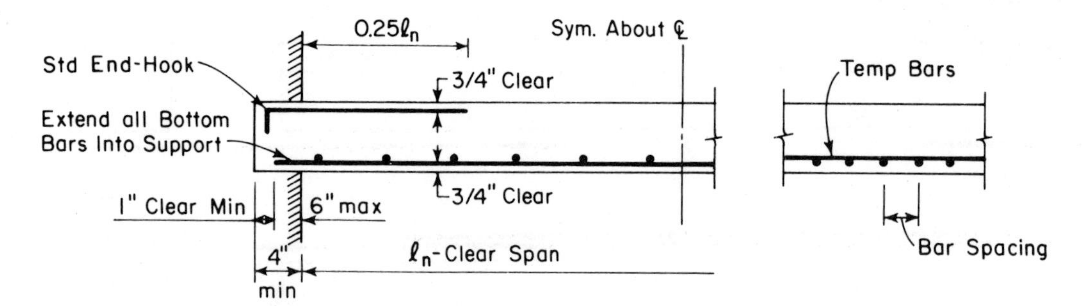

SINGLE SPAN-SIMPLY SUPPORTED

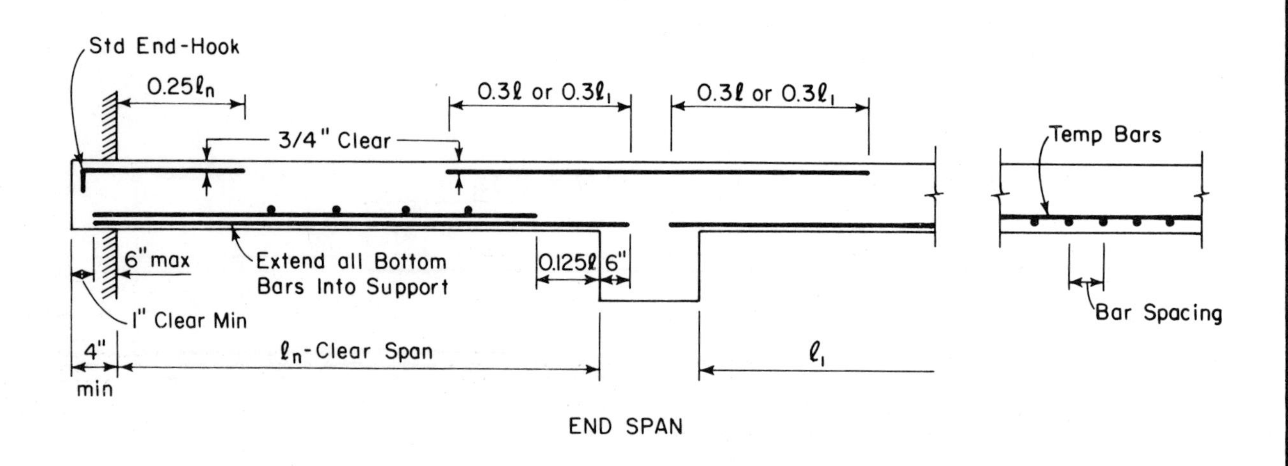

END SPAN

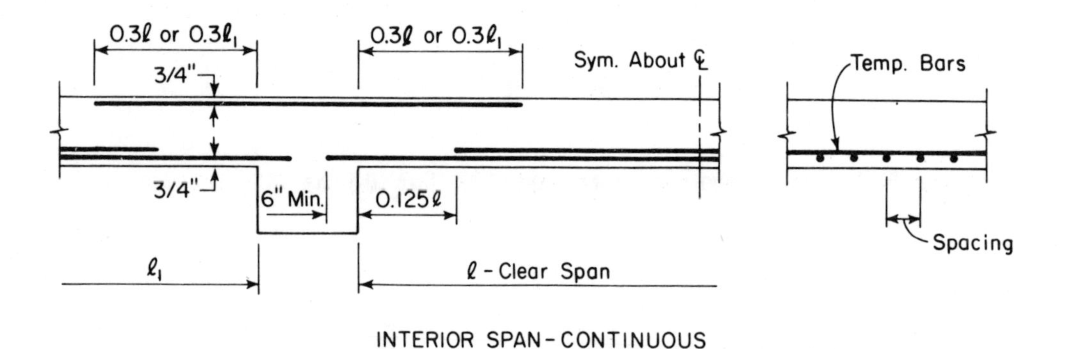

INTERIOR SPAN-CONTINUOUS

Figure 8-2 - Recommended Bar Details for One-Way Slabs

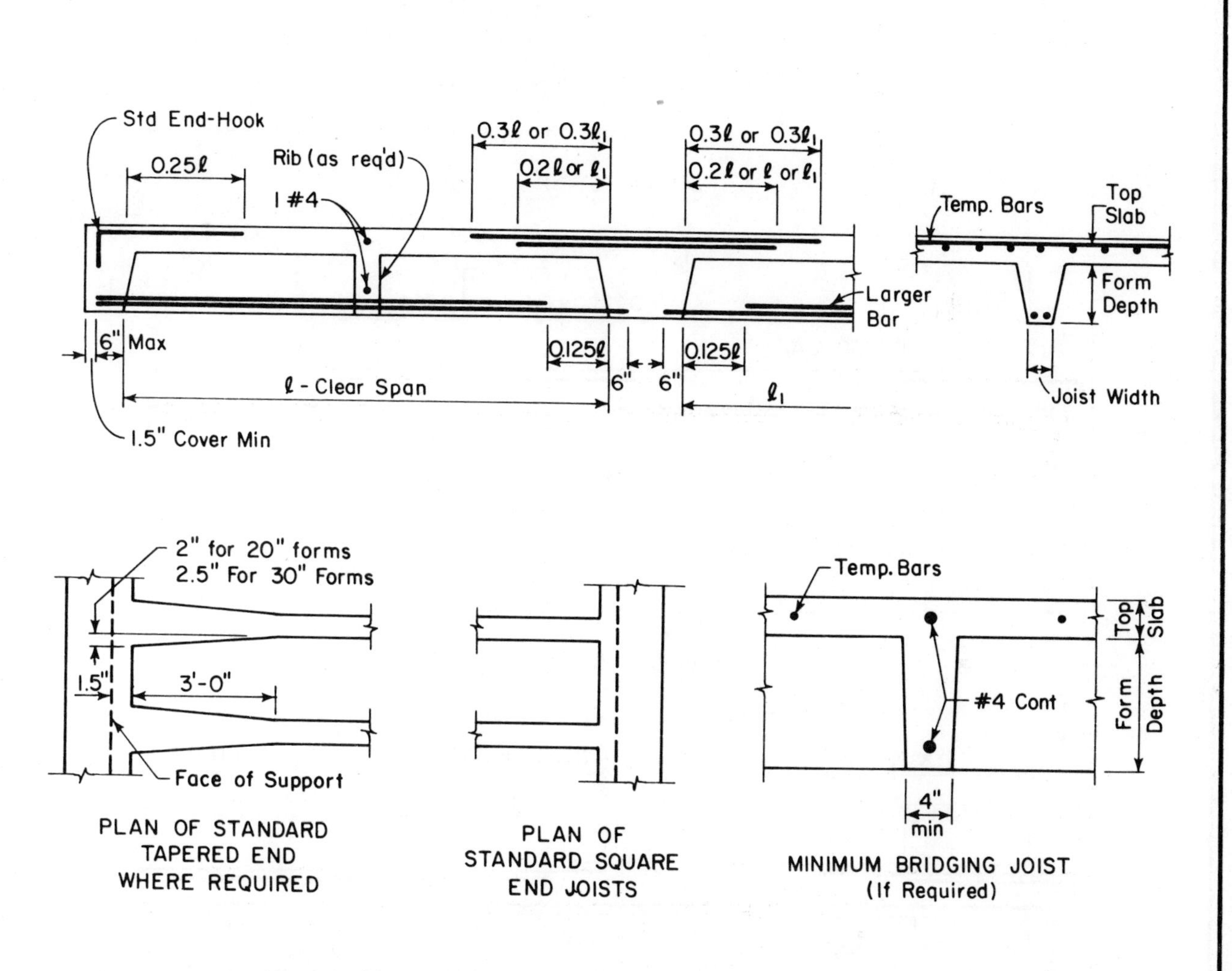

Figure 8-3 - Recommended Bar Details for One-Way Joist Construction

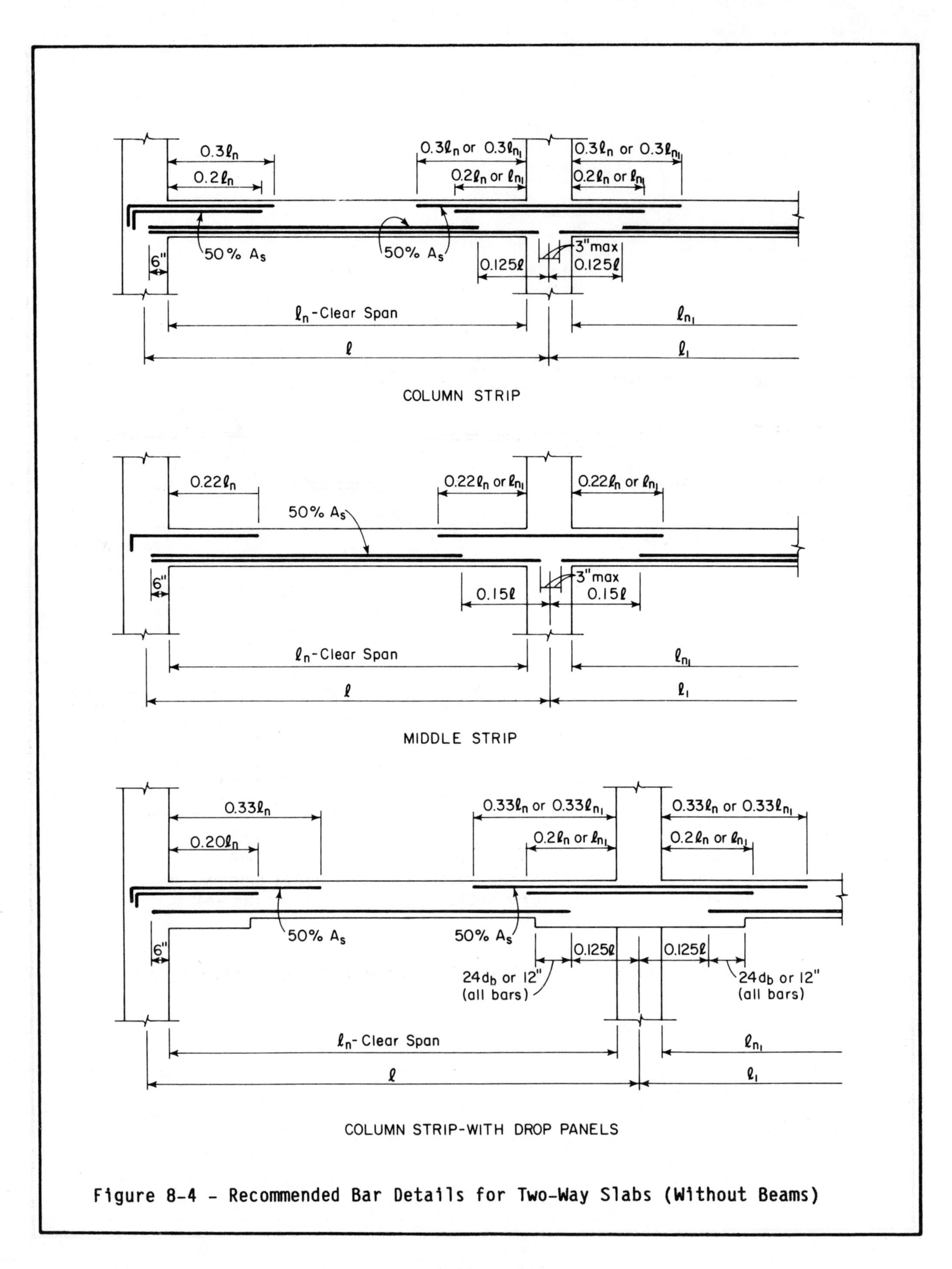

Figure 8-4 - Recommended Bar Details for Two-Way Slabs (Without Beams)

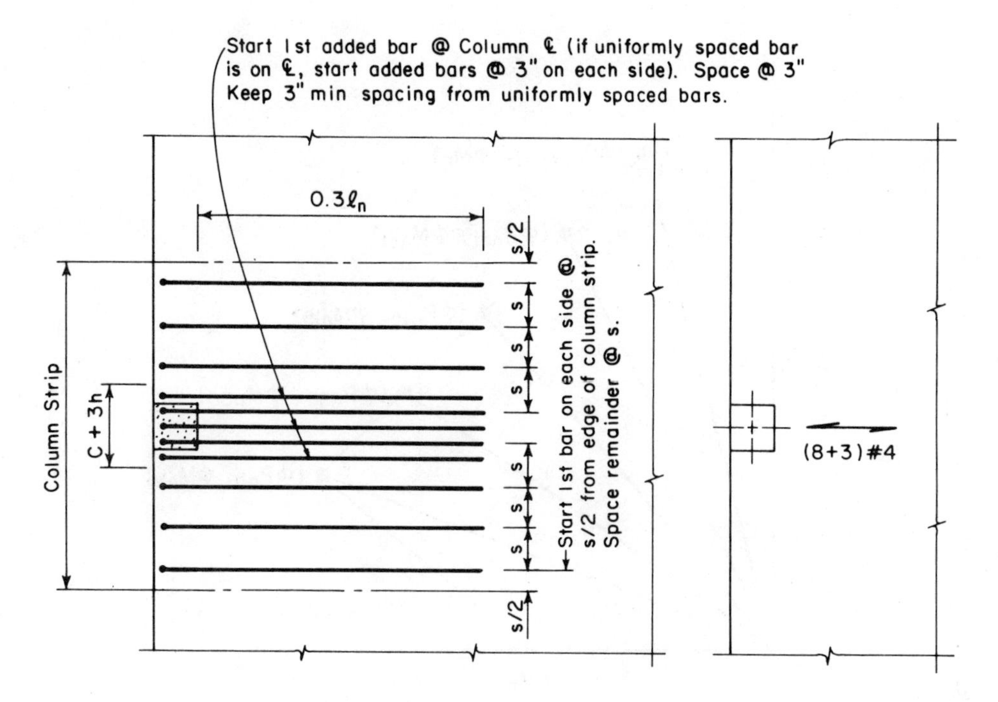

Notes: •- Spacing s = 2 x slab thickness ≤ 18 in.
•- Where added top bars are required, total bars required are shown on design drawings thus; (8+3) #4 to indicate:
8 - Uniformly spaced top bars
3 - Added bars

Fig. 8-5 - Typical Details - Top Bars at Edge Columns

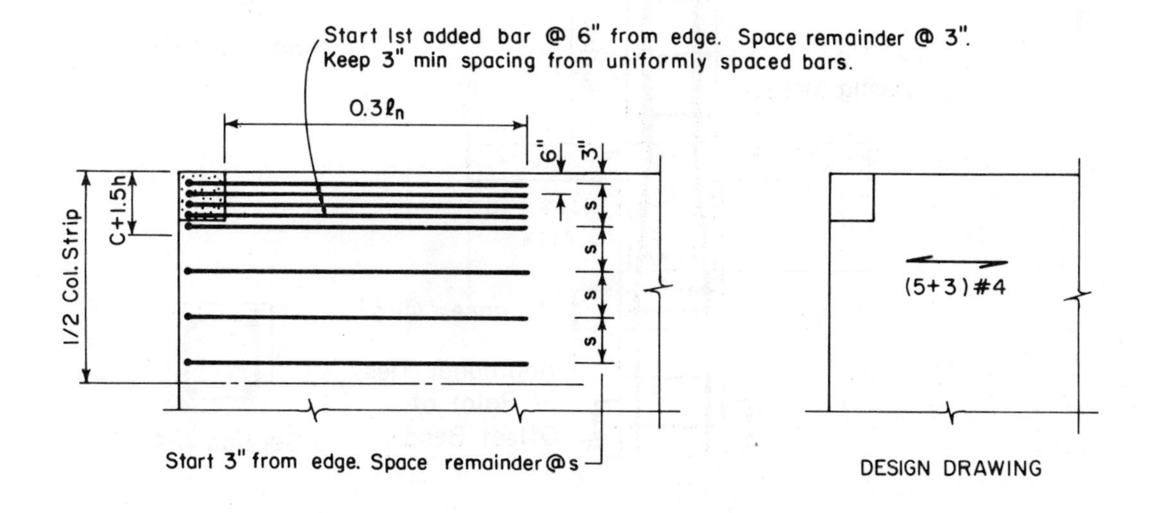

Figure 8-6 - Typical Details - Top Bars at Corner Columns

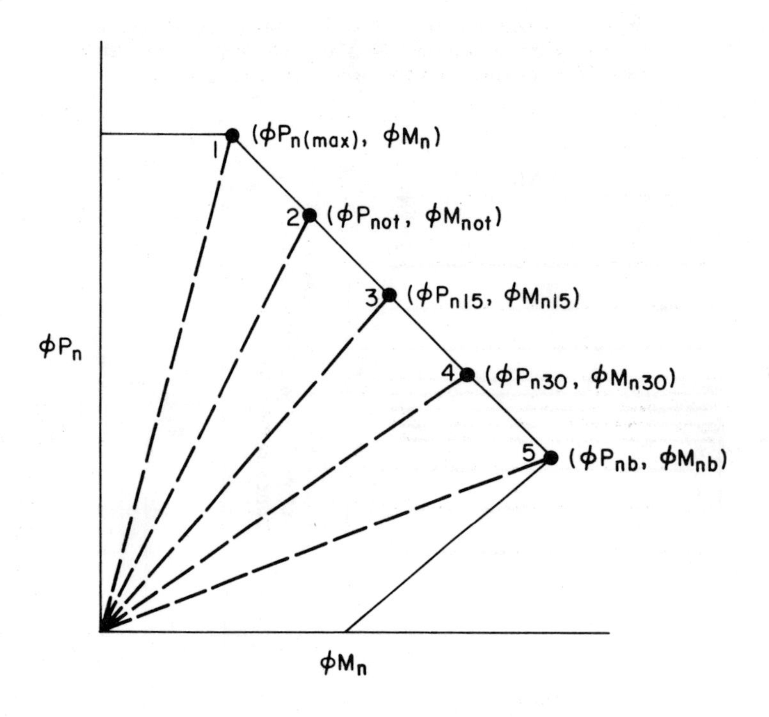

Fig. 8-7 - Load-Moment Strength Diagram

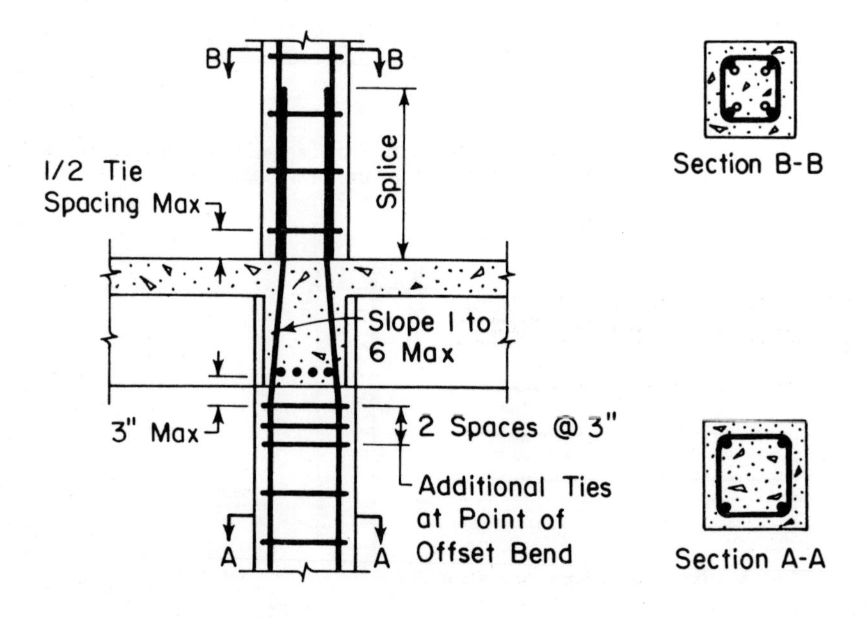

Fig. 8-8 - Column Splice Details

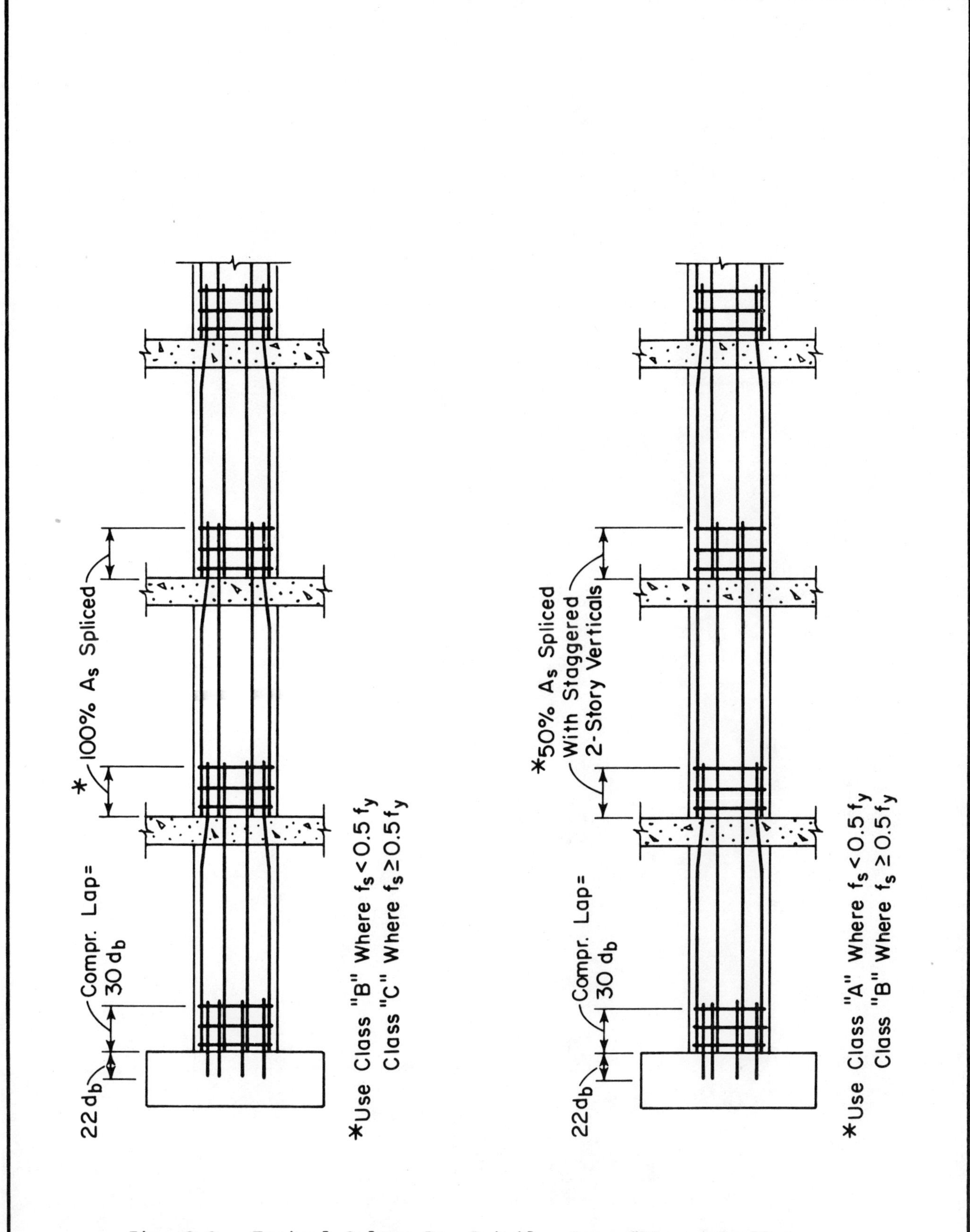

Fig. 8-9 - Typical Column Bar Details, Bars #11 and Smaller

9

Design Considerations for Economical Formwork

William R. Anthony*

9.1 - FORMWORK....The Economic Paradox

While this book is primarily dedicated to presenting timesaving ideas in the design of cast-in-place reinforced concrete structures, part of the total outlook for simplifying or making effective use of ACI 318 provisions also relates to promoting economies in construction. Since formwork can constitute as high as 50 percent of the total cost of a concrete structural frame, it is one of the logical places to look for economy.

Formwork economies should first be considered at the conceptual stage or the preliminary design phase of a construction project; a time when building load requirements are established and plan layouts and bay spacings are set; a time when the aesthetic characteristics and structural systems are determined; and a time when the mechanical and electrical features are decided. Indeed, it is that time in the course of the construction project when the entire architectural, structural, mechanical, and electrical building composite is conceived and initially developed.

Striving for construction economy is a familiar story. Design professionals, after having considered several alternate structural framing systems of cast-

*Manager, Market Development, Concrete Construction Division, The CECO Corporation

in-place concrete and having determined those systems that best satisfy loading requirements as well as other design criteria, often make their final selections on the system that would render the least use of concrete material and possibly the least quantity of reinforcing steel. This kind of approach to economical concrete construction can sometimes result in a costly design. Complex structural frames and nonstandard configurations in concrete cross sections can complicate construction to the extent that any cost savings to be realized from the economical use of in-place (permanent) materials can be significantly offset by the higher costs of formwork. The story here is that in conducting cost evaluations of concrete structural frames, it is necessary to include the costs of formwork in order to determine the most economical system.

An often-stated axiom for formwork is -- Simplicity Yields Savings: Complexity Creates Cost.

9.2 - ACHIEVING FORMWORK ECONOMY....A Design Concept

Inherent within every structural frame is the opportunity to economize. And, relative to the costs of other concrete frame structure components, the high cost of formwork makes it an obvious target for closer examination. Following are three basic design principles that govern formwork economy for all site-cast concrete. Consider each as it relates to the design of current or past projects.

9.2.1 - Stay With the Standards

Because most projects do not have the budget to accommodate custom forms, basing the design on readily available standard form sizes is essential to the achievement of formwork economy. Designing for actual dimensions of standard nominal lumber is another significant cost-cutter. Finesse carpentry carries a high price. On the other hand, a simplified approach to formwork carpentry means less sawing, less piecing together, less waste, and less time. This higher level of efficiency incorporates reduced labor and material costs and fewer opportunities for error by construction workers.

9.2.2 - Repetition

Whenever possible, repeat the use of standard materials and standard forms from bay to bay of each floor, and from floor to floor to roof. This "production line" principle is critical to the achievement of maximum formwork economy. The relationship between cost and changes in depth of horizontal construction is a major design consideration. By standardizing size or, next best, varying the width, not depth, of supporting beams most requirements can be met at lowered cost because forms can be reused for all floors. To accommodate load and span variations, increase amount of reinforcement. Also, experience has shown that changing the depth of the concrete joist system from floor to roof because of the usual differences in superimposed loads actually results in added cost. Selection of individual joist depths and beam sizes might achieve minor savings in materials, but specifying a uniform depth will achieve major savings in forming costs.

9.2.3 - Anticomplexity

Expressing his preference for a crisp, uncluttered approach to architecture, the internationally known architect, Mies van der Rohe, said "less is more." In other words, the more you shed, the purer the form. As it applies to formwork, the concept of "anticomplexity" has much the same meaning. Anticomplexity is not a technique for achieving economy; rather, it is a design approach. The approach is a discriminating attention to dimensions and details.

Consider the countless variables that must be evaluated, then integrated into final project drawings. Traditionally, economy has meant a time-consuming search for ways to cut back on quantity of materials. Unfortunately, this cost-cutting measure often creates more cost--quite the opposite effect of that intended.

Repeating the economic axiom for formwork," simplicity yields savings, complexity creates cost," consider the following cost trade-offs:

Will custom forms be cost-effective? Typically, designing for standard forms is less labor intensive and incurs less cost in materials. However, if required in a quantity that allows mass production, custom forms can be as cost-effective as standard forms.

Are deep beams cost-effective? As a rule, changing the beam depth to accommodate a difference in load will result in materials savings, but can add considerably to forming costs due to field crew disruptions and increased potential for field error. Wide flat beams are more cost-effective than deep narrow beams.

Should beam and joist spacing be uniform or vary with load? Once again, many different spacings (closer together for heavy loads, farther apart for light) can mean savings in materials. Yet, the disruption in work and added labor costs required to form the variations may far exceed savings in materials.

Should column size vary with height and loading? Consistency in column size generally pays off in reduced labor costs, particularly in buildings of moderate height. Under some conditions, however, column variance will yield savings in materials that justify the increased labor costs required to form.

Are formed surface tolerances reasonable? Section 3.3.8 of ACI Standard 347[9.1] provides a way of quantitatively indicating tolerances for surface variations due to forming quality. The suggested tolerances for formed surfaces cast-in-place are shown in Table 9-1 (Table 3.3.8 of ACI 347). The following simplified guidelines on Class of Surface to be specified will minimize costs of surface finish. Specify Class 'A' finish only for critically exposed surfaces; Class 'B' for less critical, but exposed to public view; Class 'C' for all noncritical or unexposed surfaces. Use Class 'D' for unformed surfaces only. If a more stringent class of surface is specified than is necessary for a particular formed surface, the increase in cost may become disproportionate to the increase in quality as illustrated in Fig. 9-1.

Table 9-1 Permitted Irregularities in Formed Surfaces[9.1]

Type of irregularity	Class of surface			
	A	B	C	D
Gradual[1]	1/8 in.	1/4 in.	1/2 in.	1 in.
Abrupt[2]	1/8 in.	1/4 in.	1/4 in.	1 in.

1) Warping, unplaneness, and similar uniform variations from planeness measured per 5-ft template length (straightedge) placed anywhere on the surface in any direction.

2) Offsets resulting from displaced, mismatched, or misplaced forms, sheathing, or liners or from defects in forming materials.

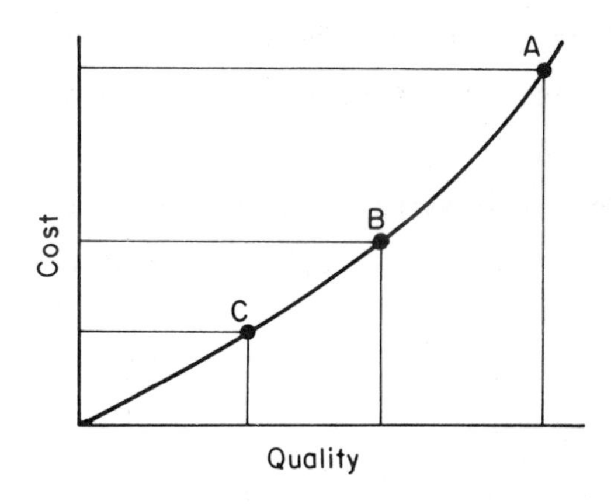

Fig. 9-1 Class of Surface vs. Cost

9.3 - HORIZONTAL FRAMING....Strategy for Economy

Floors and the required forming are usually the largest cost component of a concrete frame. The first step towards achieving maximum horizontal framing economy is selecting the system that most economically meets all load requirements.

The following framing systems are listed in ascending order of absolute cost-intensity for a given load and span size. Although beam and slab types of construction are positioned as the most expensive systems, given extreme load and span conditions, they can indeed be the most cost-effective to use.

- One-way flat slab
- Two-way flat plate
- Two-way flat slab with drops
- One-way joist slab
- Two-way joist slab (waffle)
- One-way beam and slab
- Two-way beam and slab

If bay size and load have been established, the variable then becomes cost-efficiency. Reference 9.2 provides material and cost estimating data for concrete floor and roof systems. For the design professional, the relationship between span length, floor system, and cost may indicate one or more systems to be most economical for a given project. If the system choices are equally cost-effective, then other considerations (electrical, mechanical, architectural, aesthetic, etc.) may become the determining factor.

Beyond selection of the most economical system for load and span conditions, there are general techniques that facilitate the most economical use of the chosen system.

9.3.1 - All Flat Systems

Whenever possible, avoid offsets and irregularities that cause a "stop and start" disruption of labor and require additional cutting (and waste) of materials. Steer away from breaks in soffit elevation. Depressions for terrazzo, tile, etc. should be accomplished by varying the top slab surface rather than use of offsets in the bottom of the slab. Cross section (a) is less costly to form than cross section (b).

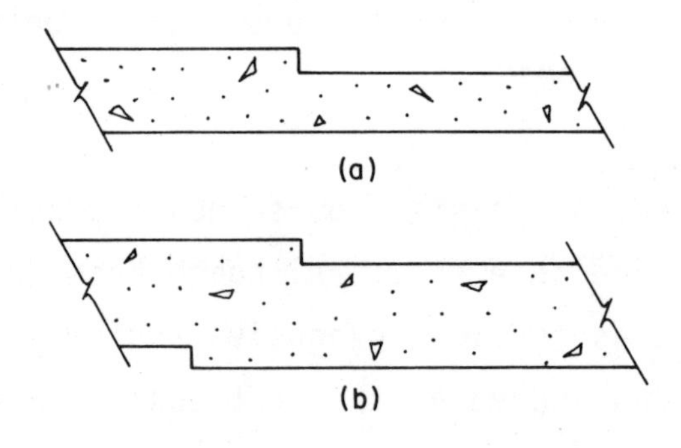

If drop panels are used at column heads, dimension drops (h_1) to equal actual nominal lumber dimension plus 3/4 in. for plyform.

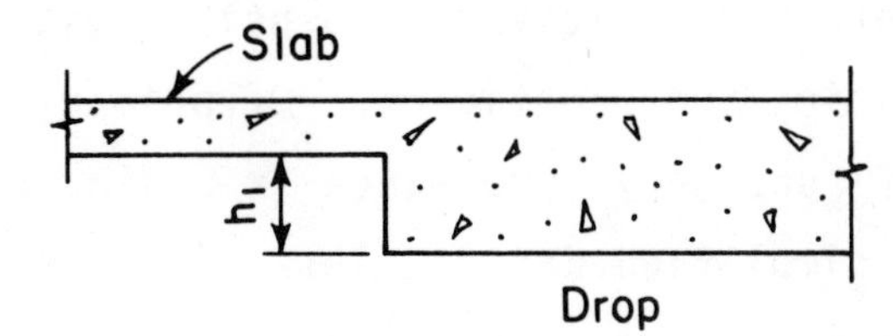

Nominal lumber dim.	Actual lumber dim.	Plyform dim.	h_1 dim.
2x	1 1/2 in.	3/4 in.	2 1/4 in.
4x	3 1/2 in.	3/4 in.	4 1/4 in.
6x	5 1/2 in.	3/4 in.	6 1/4 in.
8x	7 1/4 in.	3/4 in.	8 in.

If bay size allows, keep at least 16 ft (plus 6 in. minimum clearance) between drop panel edges. Again, this permits standard lengths of framing material to fit between drops without costly cutting of material.

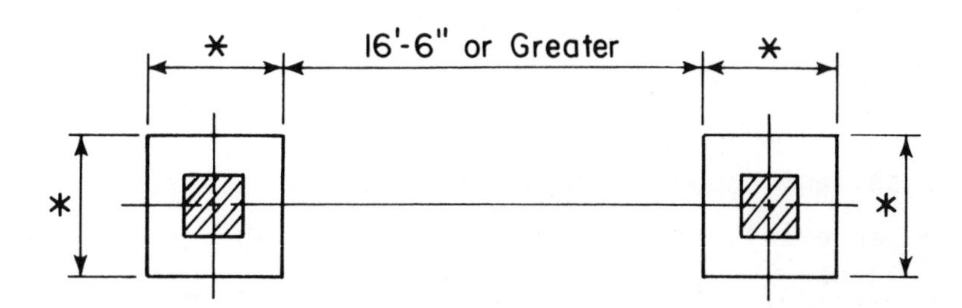

*Keep drop dimensions constant

9.3.2 - Joist Systems

Select spacing between joists that fits standard form dimensions. Whenever possible, keep joist depth consistent to avoid time-consuming field crew disruptions and its attendant costs.

Using a consistent joist width reduces costs. Variations in width mean more time for interrupted labor, more time for accurate measurement between ribs, and more opportunities for jobsite error. . . all adding to the overall cost.

9.3.3 - Beam and Slab Systems

The most economical use of this relatively expensive system relies upon the principles of standardization and repetition. Of primary importance is consistency in depth, and of secondary importance is width consistency. These two concepts will mean a simplified design, less time spent interpreting plans and more time for field crews to produce.

The second step towards achieving horizontal framing economy is to define a regular orderly progression of systematic shoring and reshoring. Avoiding high shoring conditions; specifying early form removal; and requiring a minimum of reshoring -- all indicate a concern for costs.

9.4 - VERTICAL FRAMING....Strategy for Economy

Strategically, there are three principle vertical components that should be considered when designing for maximum formwork economy:

9.4.1 - Walls

Walls provide an excellent opportunity to combine multiple functions in a single element--and an excellent opportunity to build in economy. With creative layout and design, a fire enclosure for stair or elevator shafts, columns for vertical support, and bracing can be incorporated into the same wall. Rectilinear layout and shape of walls is less costly than random patterns and cross sections.

9.4.2 - Core Areas

Core areas for elevators, stairs, and utility shafts are contained in many projects. In extreme cases, the core may require more labor per floor than the rest of the floor for an entire building. Standardizing floor openings by size and location within the core will reduce costs. Repeating the core framing pattern on as many floors as possible will also help maximize cost-efficiency.

9.4.3 - Columns

While the bulk of costs in the structural frame is contained within the floor system, column formwork economy should not be overlooked. Whenever possible, plan for uniform column locations, consistent orientations and configurations. Planning along these standard lines can yield maximum column economy as well as provide greater floor framing economy because of the uniformity in bay sizes that results.

9.5 - DESIGNER TIPS....Member Sizing for Economy

Each structural member should be examined with the concept of "keep it simple--form reuse is the key to economy."

9.5.1 - Beams

Dimensions: For a line of continuous beams, keep the beam size constant and vary the reinforcement from span to span.

Depth: Wide flat beams (same depth as slab) are easier to form than beams projecting below the bottom of the slab.

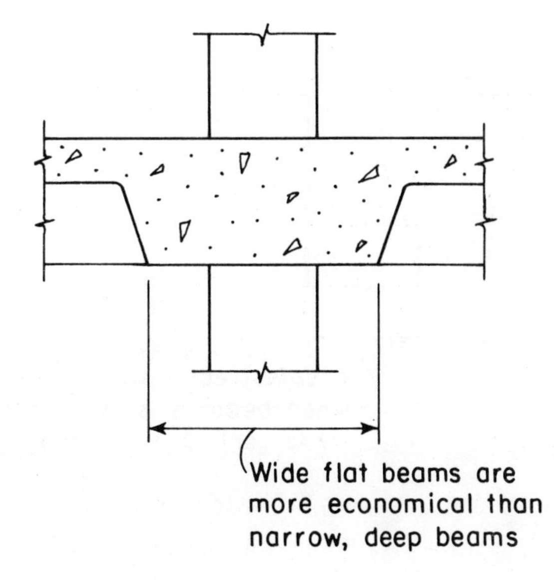

Spandrel beams are more cost intensive than interior beams due to their location at the edge of a floor slab or at a slab opening.

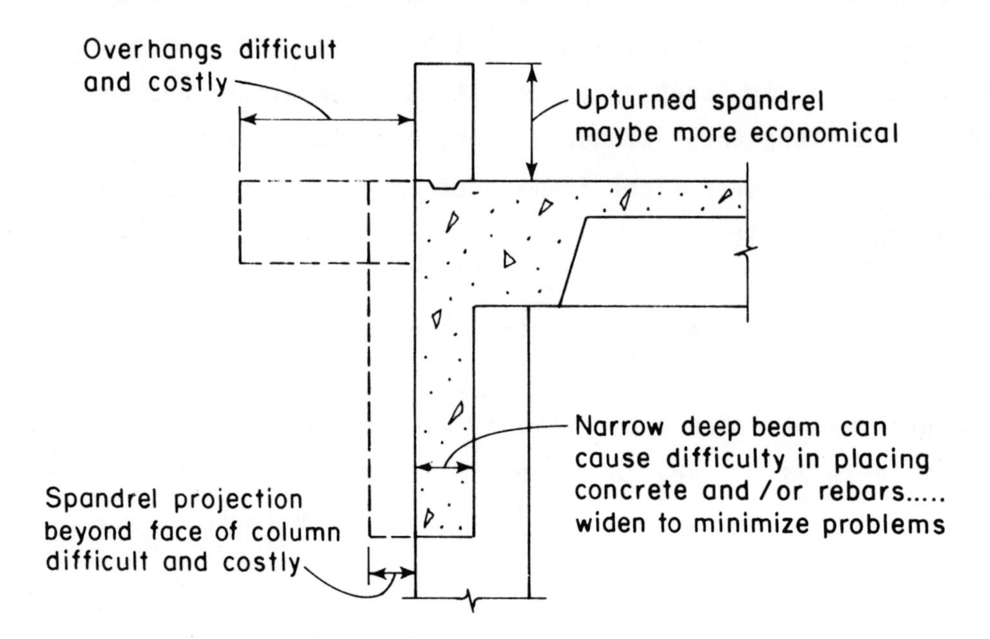

Width: Beams should be as wide as, or wider than, the columns into which they frame.

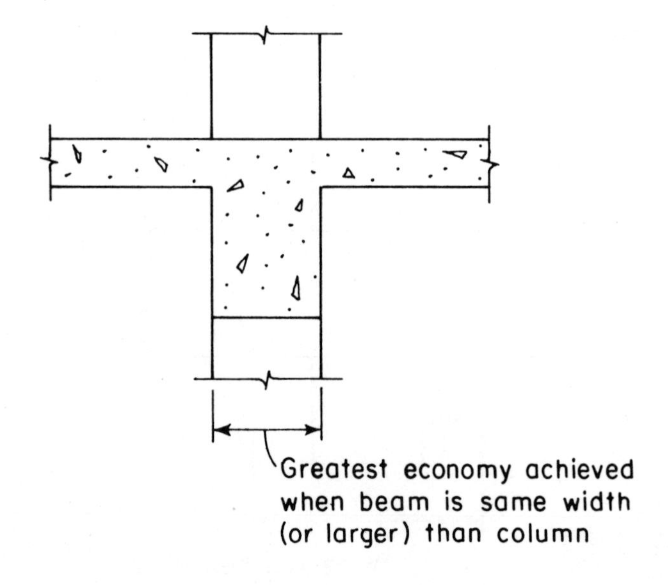

Drop Beams with Joist Systems: If loading and/or span require a drop beam, allow for minimum tee and lugs at sides of beams. Try to keep difference in elevation between bottom of beam and bottom of floor system in modular lumber dimensions.

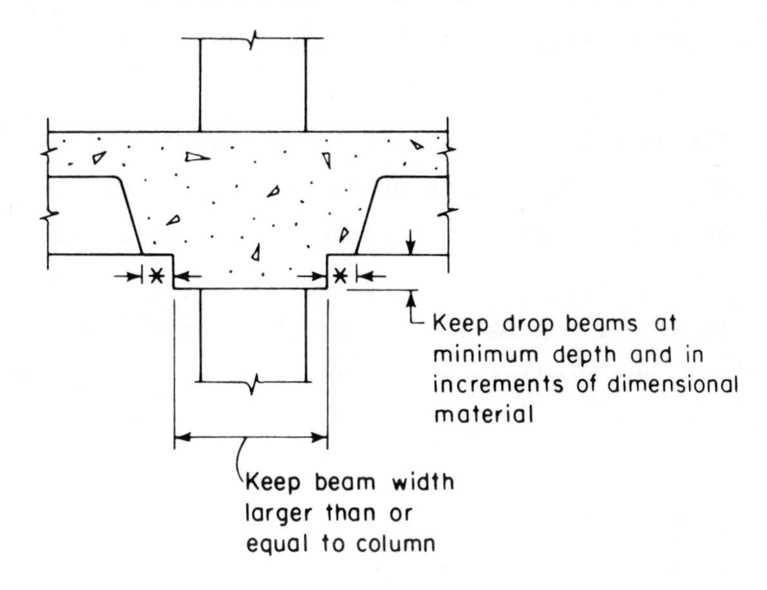

9.5.2 - Columns

Location: For maximum economy, standardize column location and orientation in a uniform pattern in both directions.

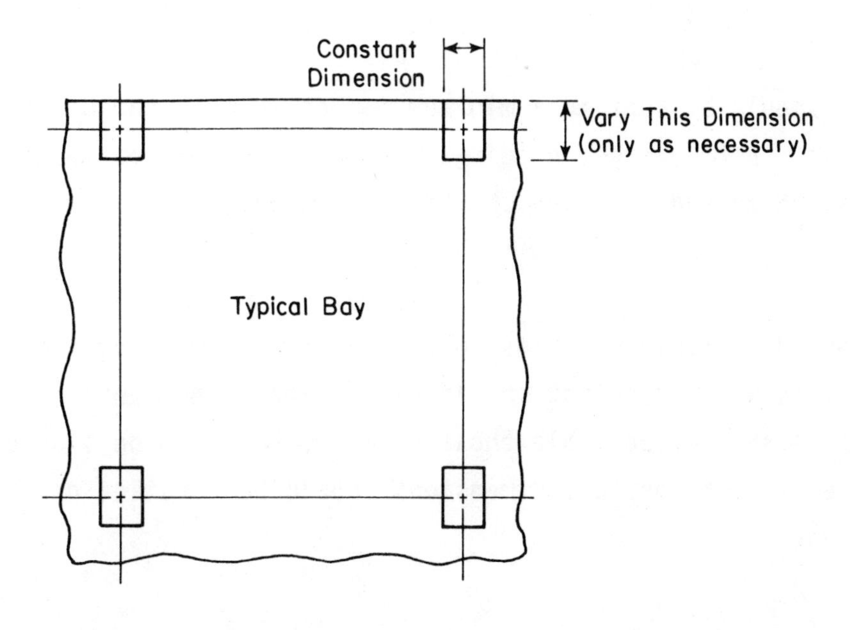

Size: Columns should be kept the same size throughout the building. If size changes are necessary, they should occur in 2 in. increments, one side at a time. (Example: 22" x 22" column should go to a 24" x 22", then 24" x 24", etc.) This approach to changing column sizes results in material economies permitting gang forming possibilities. For use with a flying form system, the distance between column faces and the flying form must be held constant. Column size changes must be made parallel to the flying form.

Shape: Use same shape as often as possible throughout any given floor and vertically from floor to floor. Square or round columns are the most economical. Use other shapes only when architectural requirements so dictate.

Column/Beam Connections: If the beam is wider than the column, the connection is simple; if not, the connection is intricate and considerably more costly.

9.5.3 - Walls

Thickness: Should be the same throughout a project if possible. Constant wall thickness facilitates the reuse of equipment, ties, and hardware. Also, this minimizes the possibilities of error in the field. In all cases, maintain sufficient wall thickness to permit proper placing and vibrating of concrete.

Openings: These should be kept to a minimum number as they can be costly and time-consuming. A few larger openings are more cost-effective than many smaller openings. Size and location should be constant for maximum reuse of formwork.

Brickledges: Should be kept at a constant height with a minimum number of steps. Thickness as well as height should be in dimensional units of lumber, approximating as closely as possible those of the masonry to be placed thereon. Brickledge locations and dimensions should be detailed on the structural drawings.

Footing Elevations: Should be kept at a constant elevation along any given wall if possible. This facilitates the use of wall gang forms from footing to footing. If footing steps are required, use minimum number possible.

Pilasters: For buildings of moderate height, pilasters can be used to transfer column loads into the foundation walls. Gang forms can be used more easily if the pilaster sides are splayed as shown below.

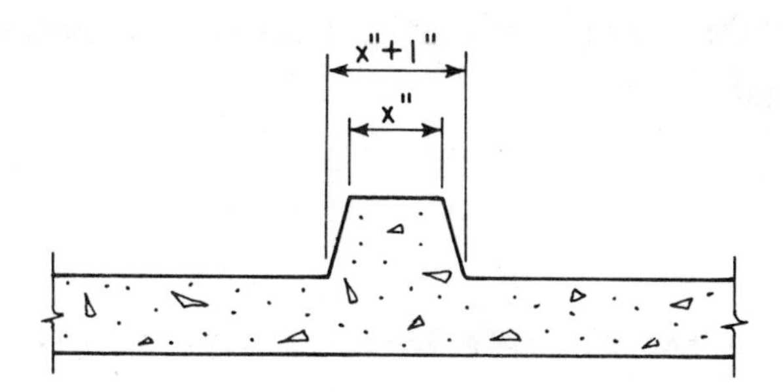

9.6 - OVERALL STRUCTURAL ECONOMY

While it has been the primary purpose of this chapter to focus on those considerations that will significantly impact the costs of the structural system relative to formwork requirements, the following are 10 general steps that can be taken during the preliminary and final design phases of the construction project that will lead to overall structural economies:

1. Study the structure as a whole.
2. Prepare freehand alternative sketches comparing all likely structural framing systems.
3. Establish column locations as uniformly as possible keeping orientation and size constant wherever possible.
4. Select preliminary sizes from concrete design aids and "rules of thumb."
5. Make cost comparisons based on sketches from Step 2 quickly, roughly, but with an adequate degree of accuracy.
6. Select the best balance between cost of structure and architectural/mechanical design considerations.
7. Distribute prints of selected framing scheme to all designer team members to reduce unnecessary future changes.

8. Plan your building. Visualize how forms would be constructed. Where possible, keep beams and columns simple without haunches, brackets, widened ends or offsets. Standardize concrete sizes for maximum reuse of forms.
9. During final design, place most emphasis on those items having greatest financial impact on total structural frame cost.
10. Plan your specifications to minimize construction costs and time by including items such as early stripping time for formwork and acceptable tolerances for finish.

Selected References

9.1 "Recommended Practice for Concrete Formwork (ACI 347-78)," American Concrete Institute, Detroit, Michigan, 1978, 37 pp.

9.2 "Concrete Floor and Roof Systems - Material and Cost Estimating Guide," Portland Cement Association, Skokie, Illinois, PA136.03B, 1982, 11 pp.

10

Design Considerations for Fire Resistance

James P. Barris*

10.1 - INTRODUCTION

State and municipal building codes throughout the country regulate the fire resistance of the various elements and assemblies comprising the building structure. Structural frames, roof/floor systems, and walls must be able to withstand the stresses and strains imposed by fully developed fires and carry their own dead loads and superimposed loads without collapse.

Fire ratings assigned to the various elements of construction are a measure of the endurance needed to safeguard the structural stability of the building during the course of a fire and to prevent the spread of fire to other parts of the building.

The determination of fire rating requirements in building codes is based on the expected fire severity associated with the type of occupancy, the amount of combustible contents (fuel loading) normally found within the occupancy classification, and the building height and area.

In the design of structures, building code provisions for fire resistance are sometimes overlooked and this may lead to costly mistakes. It is not uncommon, for instance, to find that a concrete slab in a waffle slab floor

*Director, Codes and Standards Department, PCA

system may only require a 3 to 4-1/2 in. thickness to satisfy ACI strength requirements. But, if the building code specifies a 2-hour fire resistance rating for that particular floor system, the concrete slab may need to be increased to 3-1/2 to 5 in. thickness, depending on type of concrete aggregate used. Indeed, under such circumstances and from the standpoint of economics, the fire-resistive requirements would indicate that another system of construction might be more appropriate, say, a pan-joist or flat slab floor system. Simply stated, structural members may differ significantly in their dimensional requirements between those predicated on ACI 318 strength design criteria and those necessary to attain proper fire resistance as prescribed in local building codes. Building officials are required to enforce the strictest provisions.

The purpose of this chapter is to make the reader aware of the importance of first examining the fire resistance provisions of the governing building code before proceeding with the structural design.

The field of fire technology is highly involved and complex and it is not the intent here to deal with the chemical or physical characteristics of fire, nor with the behavior of structures in real fire situations. Rather, the goal is to present some basic information as an aid to designers in establishing those fire protection features of construction that may impact their structural design work.

The information given in this chapter is fundamental. Modern day designs, however, must deal with many combinations of materials and it was not possible to handle all the intricacies of construction. Rational methods of design for dealing with more involved fire resistance problems are available. For more comprehensive discussions on the subject of the fire resistive qualities of concrete and for calculation methods used in solving design problems related to fire integrity, the reader is referred to Reference 10.1.

10.2 - DEFINITIONS

Structural Concrete:

- Siliceous aggregate concrete - concrete made with normal weight aggregates consisting mainly of silica or compounds other than calcium or magnesium carbonate.

- Carbonate aggregate concrete - concrete made with aggregates consisting mainly of calcium or magnesium carbonate, e.g., limestone or dolomite.

- Sand-lightweight concrete - concrete made with a combination of expanded clay, shale, slag, or slate or sintered fly ash and natural sand. Its unit weight is generally between 105 and 120 pcf.

- Lightweight aggregate concrete - concrete made with aggregates of expanded clay, shale, slag, or slate or sintered fly ash, and weighing 85 to 115 pcf.

Insulating Concrete:

- Cellular concrete - a lightweight insulating concrete made by mixing a preformed foam with portland cement slurry and having a dry unit weight of approximately 30 pcf.

- Perlite concrete - a lightweight insulating concrete having a dry unit weight of approximately 30 pcf made with perlite concrete aggregate produced from volcanic rock that, when heated, expands to form a glasslike material or cellular structure.

- Vermiculite concrete - a lightweight insulating concrete made with vermiculite concrete aggregate, a laminated micaceous material produced by expanding the ore at high temperatures. When added to a portland cement slurry the resulting concrete has a dry unit weight of approximately 30 pcf.

Miscellaneous Insulating Materials:

- Glass fiber board - fibrous glass roof insulation consisting of inorganic glass fibers formed into rigid boards using a binder. The board has a top surface faced with asphalt and kraft reinforced with glass fibers.

- Mineral board - a rigid felted thermal insulation board consisting of either felted mineral fiber or cellular beads of expanded aggregate formed into flat rectangular units.

10.3 - FIRE RESISTANCE RATINGS

10.3.1 - Fire Test Standards

The fire-resistive properties of building components and structural assemblies are determined by standard fire test methods. The most widely used and nationally accepted test procedure is that developed by the American Society of Testing and Materials (ASTM). It is designated as the ASTM E-119, Standard Methods of Fire Tests of Building Construction and Materials. Other accepted standards, essentially alike, include the National Fire Protection Association Standard Method No. 251; Underwriters Laboratories, U.L. 263; American National Standards Institute, No. A2-1; Underwriters Laboratories of Canada, ULC-S101; and Uniform Building Code Standard No. 43-1.

10.3.2 - ASTM E-119 Test Procedure

A standard fire test is conducted by placing the assembly in a test furnace. Floor and roof specimens are exposed to controlled fire from beneath, beams from the bottom and sides, walls from one side, and columns from all sides. The temperature is raised in the furnace over a given period of time in accordance with the ASTM E119 standard time-temperature curve shown in Fig. 10-1.

This specified time-temperature relationship provides for a furnace temperature of 1000°F at five minutes of the test, 1300°F at 10 minutes, 1700°F at one hour, 1850°F at two hours, and 2000°F at four hours. The end of the test is reached and the fire endurance of the specimen is established when any one of the following conditions first occurs:

(1) For walls, floors, and roof assemblies the temperature of the unexposed surface rises an average of 250°F above its initial temperature or 325°F at any location. In addition, walls must sustain a hose stream test.

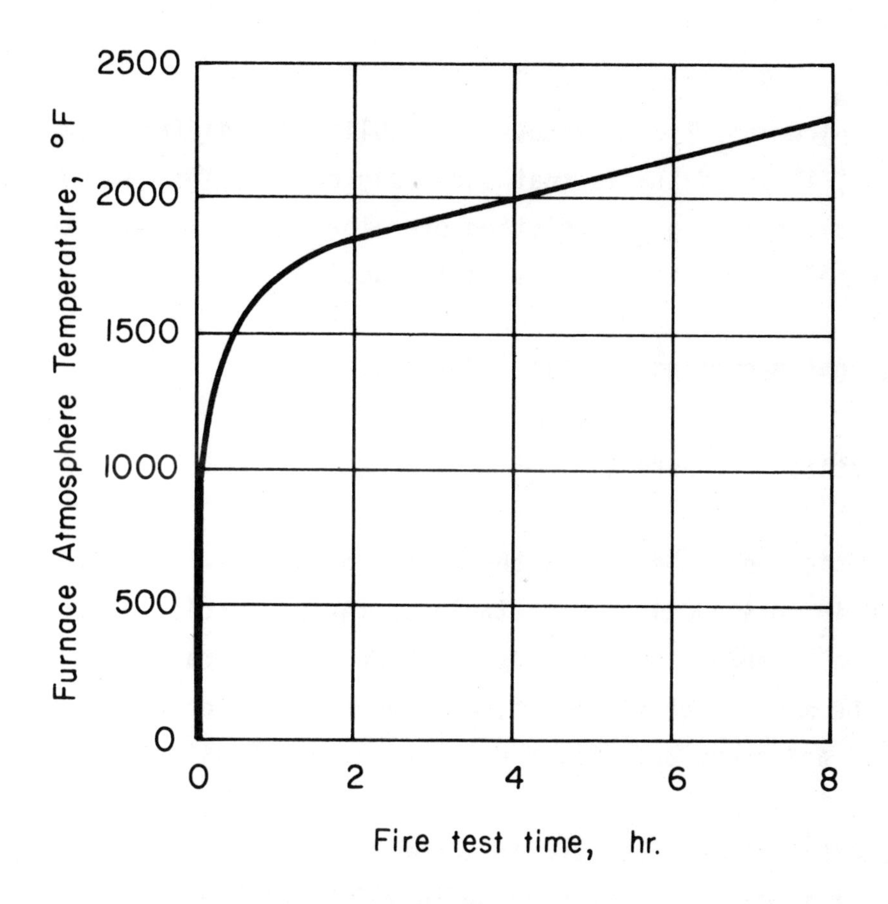

Fig. 10-1 Standard time-temperature relationship of furnace atmosphere (ASTM E119)

(2) Cotton waste placed on the unexposed side of a wall, floor, or roof system is ignited through cracks or fissures developed in the specimen.

(3) The test assembly fails to sustain the applied load. (In 1970, an additional criterion limiting the reinforcing steel temperature to 1100°F

was added for certain restrained and all unrestrained floors, roofs, and beams).

Though the complete requirements of ASTM E119 and the conditions of acceptance are much too detailed for inclusion in this chapter, experience shows that concrete floor/roof assemblies and walls will usually fail by heat transmission (item 1); and columns and beams by failure to sustain the applied loads (item 3), or beam reinforcement fails to meet the temperature rise criterion (item 3).

Fire rating requirements for structural assemblies may differ from code to code; therefore, it is advisable that the designer take into account the building regulations having jurisdiction over the construction rather than relying on general perceptions of accepted practice.

10.4 - DESIGN CONSIDERATIONS FOR FIRE RESISTANCE

10.4.1 - Properties of Concrete

Concrete is considered to be one of the most highly fire-resistive structural materials used in construction. Nonetheless, the properties of concrete and reinforcing steel change significantly in the high temperatures of fire. Strength and the modulus of elasticity are reduced, the coefficient of expansion increases, and creep and stress relaxations are considerably higher.

Concrete strength, the main concern in uncontrolled fires, remains comparatively stable at temperatures ranging up to 900°F for some concretes and 1200°F for others. Siliceous aggregate concrete, for instance, will generally maintain its original compressive strength at temperatures up to 900°F, but can lose nearly 50% of its original strength when the concrete reaches a temperature of about 1200°F. On the other hand, carbonate aggregate and sand-lightweight concretes behave more favorably in fire, their compressive strengths remaining relatively high at temperatures up to 1400°F, and diminishing rapidly thereafter. This data reflects fire test results of specimens loaded in compression to 40% of their original compressive strength.

The temperatures stated above are the internal temperatures of the concrete and are not to be confused with the heat intensity of the exposing fire. As an example, in testing a solid carbonate aggregate slab, the ASTM standard fire exposure after 1 hour will be 1700°F, while the temperatures within the test specimen will vary throughout the section: about 1225°F at 1/4 in. from the exposed surface, 950°F at 3/4 in., 800°F at 1 in., and 600°F at 1-1/2 in.; all within the limit of strength stability.

It is to be realized that the strength loss in concrete subjected to intense fire is not uniform throughout the structural member because of the time lag required for heat penetration and the resulting temperature gradients occurring within the concrete section. The total residual strength in the member will usually provide an acceptable margin of safety.

This characteristic is even more evident in massive concrete building components such as columns and girders. Beams of normal weight concrete exposed to an ASTM E-119 fire test will, at 2 hours when the exposing fire is at 1850°F, render internal temperatures of about 1200°F at 1 in. inside the beam faces and less than 1000°F at 2 in. Obviously, the dimensionally larger concrete sections found in main framing systems will suffer far less net loss in strength (measured as a percentage of total cross-sectional area) than will lighter assemblies.

Because of the variable complexities and the unknowns of dealing with the structural behavior of buildings under fire as total multidimensional systems, building codes continue to specify minimum acceptable levels of fire endurance on a component by component basis--roof/floor assemblies, walls, columns, etc. It is known, for instance, that in a multibay building, an interior bay of a cast-in-place concrete floor system subjected to fire will be restrained in its thermal expansion by the unheated surrounding construction. Such restraint increases the structural fire endurance of the exposed assembly by placing the heated concrete in compression. The restraining forces developed are large and, under elastic behavior, would cause the concrete to exceed its original compressive strength were it not for stress relaxations that occur at high temperatures.

Building codes give credit for restrained conditions through acceptance of ASTM fire test results on structural components. Tables 10-3 and 10-4 show the differences in cover requirements for reinforcement in cast-in-place concrete slabs and beams under restrained and unrestrained conditions. The use of calculation methods for determining fire endurance are also accepted, depending on local code adoptions (see Reference 10.1).

10.4.2 - Thickness Requirements

Test findings show that fire resistance in concrete structures will vary in relation to the type of aggregate used. The differences are shown in Tables 10-1 and 10-2.

Table 10-1 - Minimum Thickness for Slabs and Cast-In-Place Walls (Load-Bearing and Nonload-Bearing)

Concrete Type	Minimum Thickness (inches) for Fire Resistance Rating of -				
	1 hr	1-1/2 hr	2 hr	3 hr	4 hr
Siliceous	3.5	4.3	5.0	6.2	7.0
Carbonate	3.2	4.0	4.6	5.7	6.6
Sand-lightweight	2.7	3.3	3.8	4.6	5.4
Lightweight	2.5	3.1	3.6	4.4	5.1

Table 10-2 - Minimum Sizes of Concrete Columns

Types of Concrete	Minimum Column Dimension (inches) for Fire Resistance Rating of -				
	1 hr	1-1/2 hr	2 hr	3 hr	4 hr
Siliceous	8	8	10	12	14
Carbonate	8	8	10	12	14
Sand-lightweight	8	8	9	10.5	12

In studying the tables above it is readily apparent that there may be economic benefits to be gained from the selection of the type of concrete to be used in construction. The designer is encouraged to evaluate the alternatives.

10.4.3 - Cover Requirements

Another factor to be considered in complying with fire-resistive requirements is the minimum thickness of concrete cover for the reinforcement. The concrete protection specified in the ACI code for cast-in-place concrete will generally equal or exceed the minimum cover requirements shown in the following tables, but there are a few exceptions at the higher fire ratings and these should be noted.

CONCRETE SLABS: The minimum thickness of concrete cover to the positive moment reinforcement is given in Table 10-3 for one-way or two-way slabs with flat undersurfaces.

TABLE 10-3 - Cover Thicknesses for Reinforced Concrete Floor or Roof Slabs

Concrete Aggregate Type	Minimum Thickness of Cover (inches) for Fire Resistance Rating of -							
	Restrained*				Unrestrained*			
	1 hr	1-1/2 hr	2 hr	3 hr	1 hr	1-1/2 hr	2 hr	3 hr
Siliceous	3/4	3/4	3/4	3/4	3/4	3/4	1	1-1/4
Carbonate	3/4	3/4	3/4	3/4	3/4	3/4	3/4	1-1/4
Sand-lightweight or lightweight	3/4	3/4	3/4	3/4	3/4	3/4	3/4	1-1/4

*See Appendix A4 of the ASTM E119-79 standard for guidance on restrained and unrestrained assemblies.

BEAMS: The minimum thickness of concrete cover to the positive moment reinforcement (bottom steel) for reinforced concrete beams is shown in Table 10-4.

Table 10-4 - Cover to Main Reinforcing Bars for Reinforced Concrete Beams (Applicable to All Types of Structural Concrete)

Restrained or Unrestrained*	Beam Width, in.**	Cover Thickness (inches) for Fire Resistance Rating of -				
		1 hr	1-1/2 hr	2 hr	3 hr	4 hr
Restrained	5	3/4	3/4	3/4	1(*)	1-1/4(*)
Restrained	7	3/4	3/4	3/4	3/4	3/4
Restrained	≥ 10	3/4	3/4	3/4	3/4	3/4
Unrestrained	5	3/4	1	1-1/4	---	---
Unrestrained	7	3/4	3/4	3/4	1-3/4	3
Unrestrained	≥ 10	3/4	3/4	3/4	1	1-3/4

*See Appendix A4 of the ASTM E119-79 standard for guidance on restrained and unrestrained assemblies. Tabulated values for restrained assemblies apply to beams spaced more than 4 ft on centers; for restrained beams spaced 4 ft or less on centers, minimum cover of 3/4 in. is adequate for ratings of 4 hr or less.

**For beam widths between the tabulated values, the minimum cover thickness can be determined by direct interpolation.

COLUMNS: The minimum cover to main reinforcement in columns is shown in Table 10-5.

Table 10-5 - Cover for Reinforced Concrete Columns

Concrete Aggregate Type	Minimum Cover Thickness (inches) for Fire Resistance Rating of -				
	1 hr	1-1/2	2 hr	3 hr	4 hr
Siliceous	1-1/2	1-1/2	1-1/2	1-1/2	2
Carbonate	1-1/2	1-1/2	1-1/2	1-1/2	1-1/2
Sand-lightweight	1-1/2	1-1/2	1-1/2	1-1/2	1-1/2

10.5 - MULTICOURSE FLOORS AND ROOFS

Symbols: Carb = carbonate aggregate concrete
Sil = siliceous aggregate concrete
SLW = sand-lightweight concrete

10.5.1 - Two-Course Concrete Floors

Fig. 10-2 gives information on the fire resistance ratings of floors that consist of a base slab of concrete with a topping (overlay) of a different type of concrete.

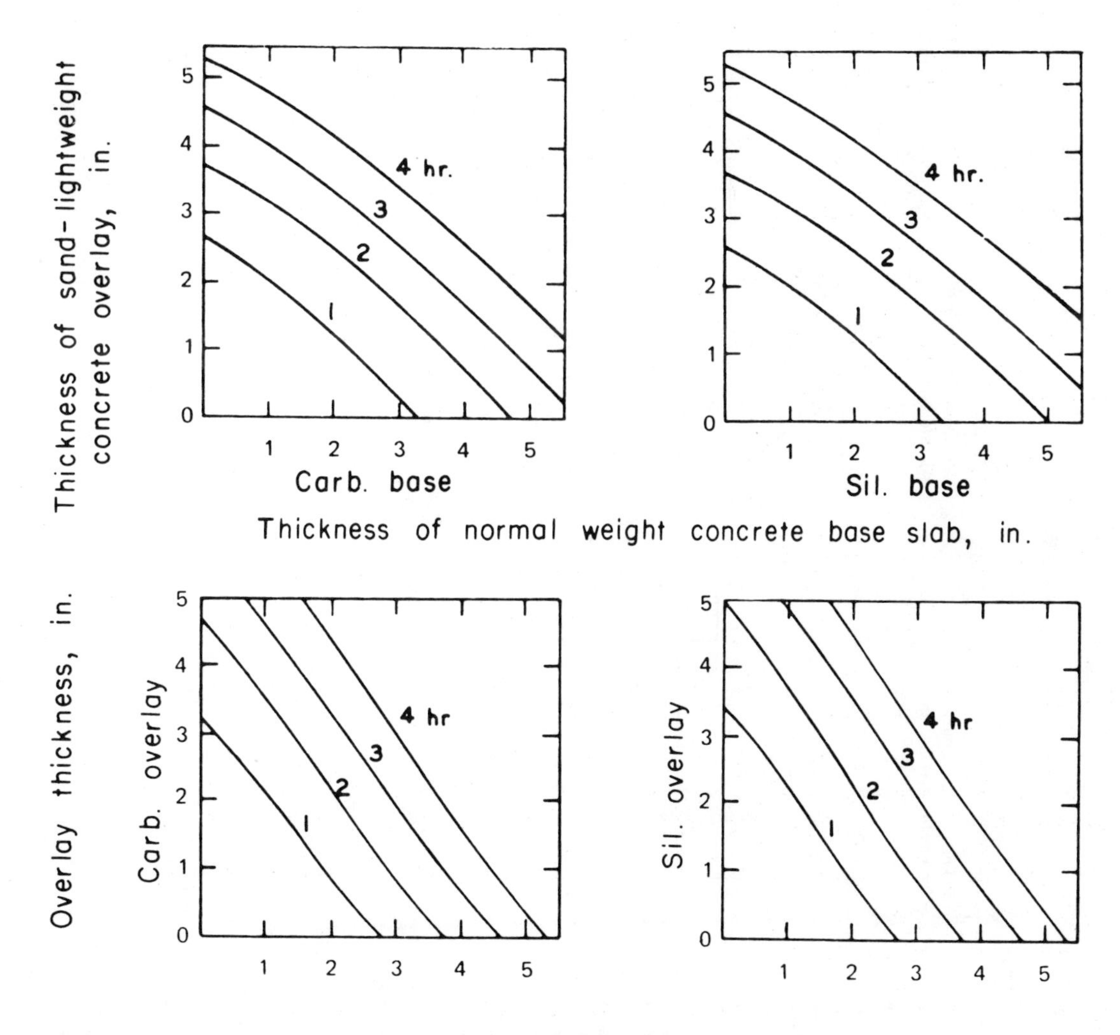

Fig. 10-2 - Fire resistance ratings for two-course floor slabs

10.5.2 - Two-Course Concrete Roofs

Fig. 10-3 gives information on the fire resistance ratings of roofs that consist of a base slab of concrete with a topping (overlay) of an insulating concrete, but not including the built-up roofing.

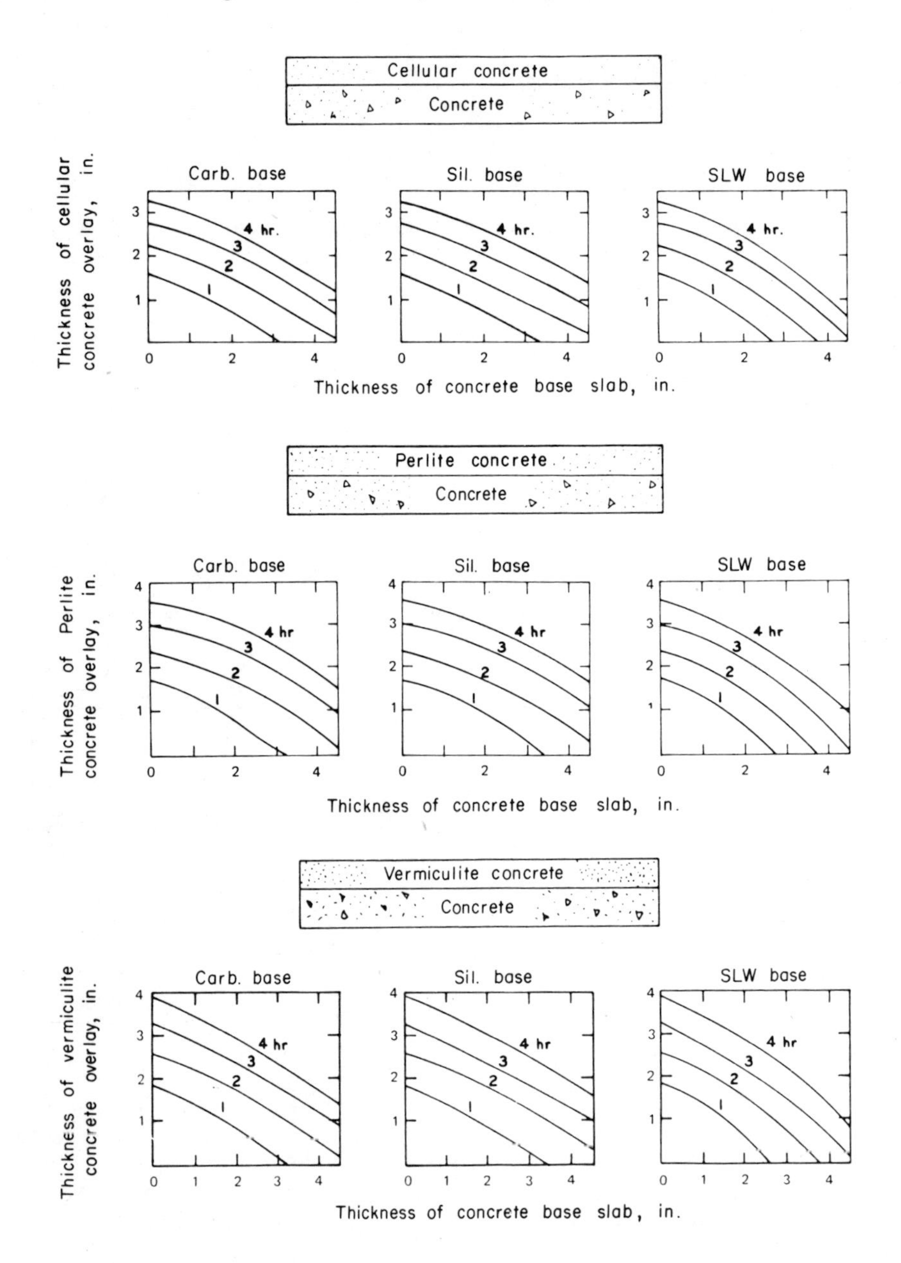

Note: For the transfer of heat, three-ply built-up roofing contributes 10 minutes to the fire resistance rating. Thus 10 minutes can be added to concrete assemblies shown above.

Fig. 10-3 - Fire resistance ratings for two-course roof slabs

10.5.3 - Concrete Roofs with Other Insulating Materials

Fig. 10-4 gives information on the fire resistance ratings of roofs that consist of a base slab of concrete with an insulating board overlay and includes the built-up roofing (three-plys).

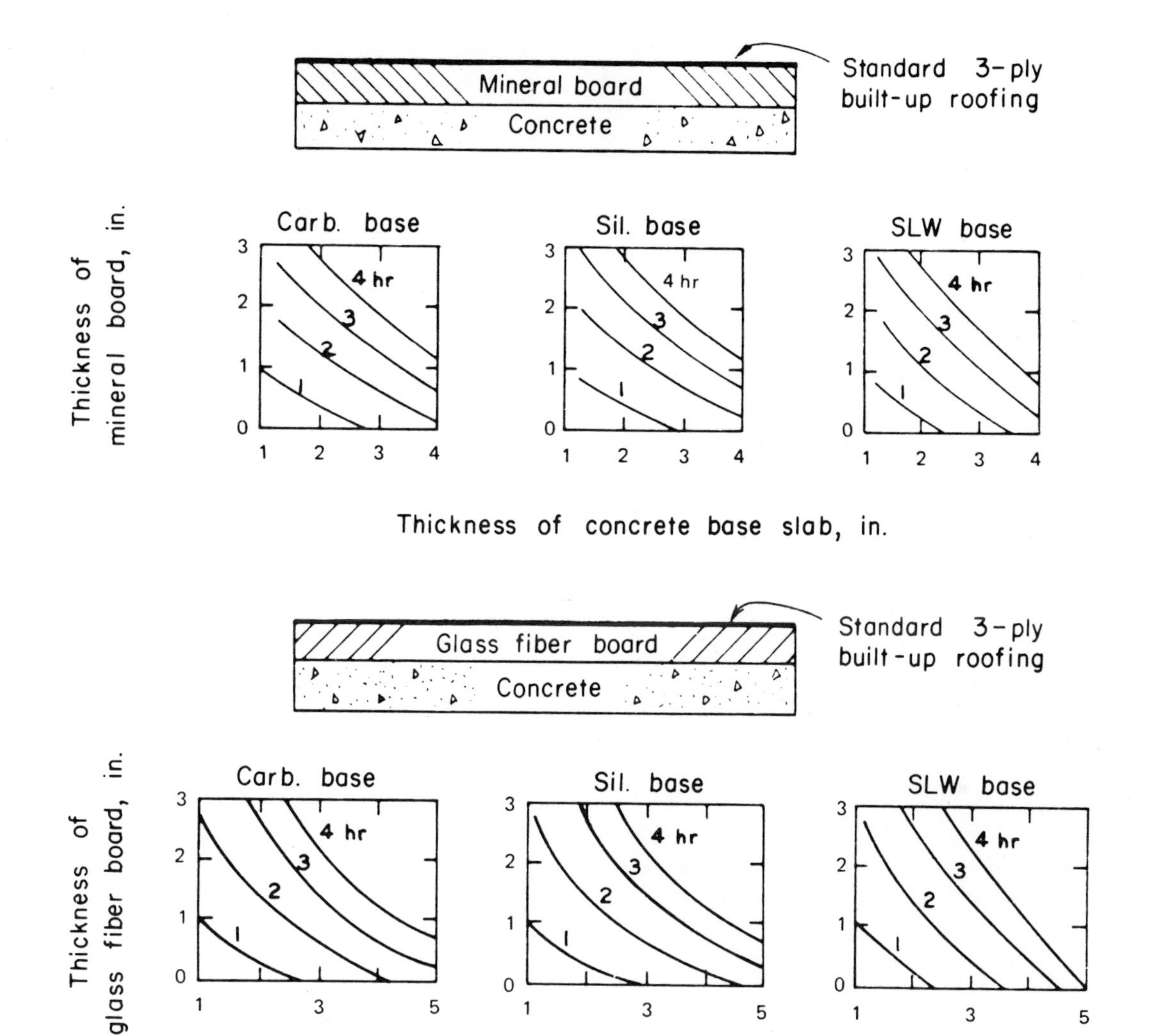

Fig. 10-4 - Fire resistance ratings for roof slabs with insulating overlays

Selected References

10.1 "Reinforced Concrete Fire Resistance," 1980 Edition, published by the Concrete Reinforcing Steel Institute (CRSI), 933 North Plum Grove Road, Schaumburg, Illinois 60195.